FAMILIES OF BIVARIATE DISTRIBUTIONS

AF598472

K. V. MARDIA

M.Sc. (Bombay), M.Sc. (Poona), Ph.D. (Rajasthan), Ph.D. (Newcastle)
Senior Lecturer in Mathematical Statistics
University of Hull

BEING NUMBER TWENTY-SEVEN OF
GRIFFIN'S STATISTICAL
MONOGRAPHS & COURSES
EDITED BY
ALAN STUART, D.Sc.(Econ.)

GRIFFIN 1820 LONDON

CHARLES GRIFFIN & COMPANY LIMITED
42 DRURY LANE, LONDON, WC2B 5RX

Copyright © K. V. MARDIA 1970
All rights reserved

No part of this publication may be reproduced, stored in a retrieval system, or transmitted, in any form or by any means, electronic, mechanical, photocopying, recording or otherwise, without the prior permission of the Publishers, as above named.

First published 1970
ISBN: 0 85264 182 6

Demy Octavo, ix + 109 pages
12 line illustrations, 5 tables

Set by E W C Wilkins & Associates Ltd London N12
Printed in Great Britain by Latimer Trend & Co Ltd Whitstable

057523

PUBLISHERS' NOTE

The series of monographs in which this title appears was introduced by the publishers in 1957, under the General Editorship of Dr Maurice G. Kendall. Since that date, more than twenty volumes have been issued, and in 1966 the Editorship passed to Alan Stuart, D.Sc.(Econ.), Professor of Statistics, University of London.

The Series fills the need for a form of publication at moderate cost which will make accessible to a group of readers specialized studies in statistics or courses on particular statistical topics. Often, a monograph on some newly developed field would be very useful, but the subject has not reached the stage where a comprehensive treatment is possible. Considerable attention has been given to the problem of producing these books speedily and economically.

It is intended that in future the Series will include works on applications of statistics in special fields of interest, as well as theoretical studies. The publishers will be interested in approaches from any authors who have work of importance suitable for the Series.

CHARLES GRIFFIN & CO. LTD.

GRIFFIN'S STATISTICAL MONOGRAPHS AND COURSES

No. 1: *The analysis of multiple time-series* M. H. QUENOUILLE
No. 2: *A course in multivariate analysis* M. G. KENDALL
No. 3: *The fundamentals of statistical reasoning* M. H. QUENOUILLE
No. 4: *Basic ideas of scientific sampling* A. STUART
No. 5: *Characteristic functions** E. LUKACS
No. 6: *An introduction to infinitely many variates* E. A. ROBINSON
No. 7: *Mathematical methods in the theory of queueing* A. Y. KHINTCHINE
No. 8: *A course in the geometry of* n *dimensions* M. G. KENDALL
No. 9: *Random wavelets and cybernetic systems* E. A. ROBINSON
No. 10: *Geometrical probability* M. G. KENDALL and P. A. P. MORAN
No. 11: *An introduction to symbolic programming* P. WEGNER
No. 12: *The method of paired comparisons* H. A. DAVID
No. 13: *Statistical assessment of the life characteristic: a bibliographic guide* W. R. BUCKLAND
No. 14: *Applications of characteristic functions* E. LUKACS and R. G. LAHA
No. 15: *Elements of linear programming with economic applications* R. C. GEARY and M. D. MCCARTHY
No. 16: *Inequalities on distribution functions* H. J. GODWIN
No. 17: *Green's function methods in probability theory* J. KEILSON
No. 18: *The analysis of variance* A. HUITSON
No. 19: *The linear hypothesis: a general theory* G. A. F. SEBER
No. 20: *Econometric techniques and problems* C. E. V. LESER
No. 21: *Stochastically dependent equations: an introductory text for econometricians* P. R. FISK
No. 22: *Patterns and configurations in finite spaces* S. VAJDA
No. 23: *The mathematics of experimental design: incomplete block designs and Latin squares* S. VAJDA
No. 24: *Cumulative sum tests: theory and practice* C. S. VAN DOBBEN DE BRUYN
No. 25: *Statistical models and their experimental application* P. OTTESTAD
No. 26: *Statistical tolerance regions: classical and Bayesian* I. GUTTMAN
No. 27: *Families of bivariate distributions* K. V. MARDIA
No. 28: *Generalized inverse matrices with applications to statistics* R. M. PRINGLE and A. A. RAYNER
No. 29: *The generation of random variates* T. G. NEWMAN and P. L. ODELL

*Now published independently of the Series.

For a list of other statistical and mathematical books see back cover.

PREFACE

Ever since the first statistical use of the bivariate normal distribution by Galton and Dickson, attempts have been made to develop families of bivariate distributions which describe non-normal variations. Earlier solutions were examined to assess their adequacy for fitting observed distributions, but the results were not encouraging. On the other hand, partly because of the relative simplicity of mathematical treatment of the bivariate normal distribution, the development of normal theory has been intensive and most thinking has centred upon it. Even nowadays, the bivariate normal distribution plays a dominant role. Recently, however, other bivariate distributions have started appearing in probabilistic models and statistical sampling problems. In addition, interest in families of bivariate distributions has been revived by certain problems arising in statistical procedures.

The literature on the subject is widely dispersed throughout the statistical journals. In this monograph, an attempt has been made to give a systematic account of the subject. It is hoped that this work will stimulate further research in the field.

It is assumed that the reader is familiar with standard univariate distributions and families of univariate distributions. Knowledge of Chapters 5 and 6 of *The Advanced Theory of Statistics* by M.G. Kendall and A. Stuart, Vol. 1 (3rd edition, 1969), is sufficient for this purpose.

Chapter 1 gives a historical background to the subject and introduces notation and terminology to be used subsequently. Chapters 2 to 8 inclusive deal with various families grouped according to the method of construction used; the order of these chapters corresponds to the order in which the underlying methods were developed. Chapter 9 gives an account of some other families of interest which cannot be fitted into the pattern of the preceding chapters. Chapter 10 deals with bivariate distributions which can be obtained by extending special features of well-known models. These distributions have been referred to in the earlier chapters and, for ease of reference, they are listed in the Appendix. The bibliography is fairly exhaustive and also serves as an author index.

The idea of this monograph originated in discussion with Professor A. Stuart who has guided me considerably in its preparation. I wish to express my deepest gratitude to him for his valuable assistance. It is also a special pleasure to thank Dr. E.A. Evans for helpful comments. Likewise I am grateful to Professor T. Lewis and to Professor R.L. Plackett for their interest and encouragement; to Professor E.S. Pearson for clarifying certain historical points in Chapter 2; and to Dr. J.W. Thompson for some stimulating discussions. Finally my thanks go to Miss T. Blackmore who so efficiently typed the difficult manuscript.

K.V.M.

HULL,
May, 1970

CONTENTS

CHAPTER 1

INTRODUCTION

1.1 Historical background

In the univariate case, various general families have been developed which will satisfactorily describe the frequency curves met with in practice. Besides this classical use of families in fitting a curve to data, the families have proved useful (i) in approximating the sampling distributions of statistics for which only the moments are readily expressible, and (ii) in providing typical forms of alternatives to explore the robustness of standard tests. For continuous random variables, some of the standard families are Karl Pearson's differential equations, the Gram–Charlier and Edgeworth expansion, and Johnson's translation system. These methods of approach can be classified as

(a) the solutions of a differential equation;

(b) series expansion using the normal distribution as a generating function;

(c) the method of translation, again usually applied to the normal distribution.

For the discrete case, Ord (1967a,b) has recently developed a family using the difference equation analogous to Pearson's differential equation.

After the development of satisfactory systems for use in the univariate case, it was only natural to attempt to extend them. Since the first statistical treatment of the bivariate normal distribution by Galton and Dickson (1886), the need for bivariate distributions to describe non-normal variations has been felt. However, the number of additional parameters then increases rapidly as we have to deal with not only the marginal distributions but also with the conditional distributions. Consequently, the problem in this case is much more intricate. The methods of approach (a)–(c) were the first to be extended to the bivariate case. Karl Pearson extended his system as the solution of two differential equations, but efforts starting in the 1890's to obtain a general solution were unsuccessful. The special solutions obtained from this approach give arbitrary restrictions on parameters. Series expansions using bivariate distributions as generating functions were also developed, but, as in the univariate case, the bivariate Edgeworth expansion suffers from the existence of negative frequencies. The translation systems, most often using a bivariate normal distribution, were considered, but each system requires only one parameter to describe dependence between the variables, so that this approach can hardly be regarded as satisfactory for purposes of fitting. Subsequently, various other models were proposed which used the inherent characteristics of bivariate distributions such as dependence. However, the possible solutions have been so many and so varied that by the 1930's the field had become intractable. It was Pretorius who first devoted himself to the task, and in his brilliant memoir of 1930 he surveyed the systems then available and critically examined their adequacy in fitting observed distributions. In fact, six observed distributions ranging from moderately skew correlation to considerably skew correlation were selected for the analysis. He concluded his long memoir with the following comment :–

> It can still be said, in conclusion, that after more than thirty years the problem still remains the "most urgent task before mathematical statisticians". The solutions that have been put forward will be serviceable in certain special cases; but no satisfactory solution to the general problem has yet been reached.

The same conclusion still applies, from the fitting point of view, to the systems so far known. But since then, several bivariate distributions have appeared in probabilistic models and in statistical sampling problems. Further, interest in families has been revived in examining the robustness of standard tests for the normal population and in investigating the efficiency of non-parametric procedures relative to their parametric competitors. There has been definite progress in this direction. E.S. Pearson (1962) has given a survey of some of the work, and a fairly complete list of distributions published before 1958 has been given by Haight (1961).

In this monograph we give a systematic account of various families. We examine the motivation behind each family and consider its outstanding features; we investigate its solutions and examine whether it contains standard distributions, a procedure which gives some idea of how elastic a family really is. We give a method of fitting the system, and the method is either illustrated or references are provided to work in which illustrations have appeared. Further, we give an account of other uses of the families, such as in studies of robustness, in non-parametric tests, in sequential analysis, and in solving some difficult problems concerned with the bivariate normal distribution. Interrelations between systems are also considered.

1.2 Notation and terminology

Let X and Y be two random variables (r.v's) with distribution function (d.f.) $H(x, y)$. The marginal d.f's of X and Y will be denoted by $F(x)$ and $G(y)$, respectively. If X and Y are continuous r.v's, we denote the joint probability density function (p.d.f.) by $h(x, y)$, the marginal p.d.f. of X by $f(x)$ and of Y by $g(y)$. The conditional p.d.f. of Y given $X = x$ will be denoted by $k(y|x)$ and the corresponding d.f. by $K(y|x)$. The conditional distribution may be described as an array distribution. For a discrete r.v. (X, Y), we shall write the probability of $X = x$ and $Y = y$ as $\mathrm{P}(X = x, Y = y)$ or briefly as $p(x, y)$. The r.v. (X, Y) degenerates on a curve $\psi(x, y) = 0$ if $\mathrm{P}[\psi(X, Y) = 0] = 1$ and $\mathrm{P}(X = x, Y = y) \neq 1$, so that the whole mass of probability is concentrated on a curve but not in a single point.

We denote the moment of order (r, s), $\mathrm{E}(X^r Y^s)$, by μ'_{rs} and the central moment of order (r, s) by μ_{rs}. We write $\mu_{10} = \mu$, $\mu_{01} = \nu$, $\mu_{20} = \mathrm{var}(X) = \sigma_1^2$, and $\mu_{02} = \mathrm{var}(Y) = \sigma_2^2$. Further, $\mu_{11} = \rho\sigma_1\sigma_2 = \mathrm{cov}(X, Y)$ and $\mathrm{corr}(X, Y) = \mu_{11}/\sigma_1\sigma_2 = \rho$. We write $\beta_{10} = \mu_{30}^2/\sigma_1^6$, $\beta_{20} = \mu_{40}/\sigma_1^4$. Further, the characteristic function of the r.v. (X, Y), $\mathrm{E}\{\exp(it_1X + it_2Y)\}$, will be denoted by $\phi_{X,Y}(t_1, t_2)$. We write $\mu'_r(x) = \mathrm{E}(Y^r|X = x) = \mathrm{E}(Y^r|x)$, $\mu_r(x) = \mathrm{E}[\{Y - \mu'_1(x)\}^r|x]$. The curve $y = \mu'_1(x)$ is the regression curve of Y on X, and the curve $y = \mu_2(x)$ or $y = \mathrm{var}(Y|x)$ is the scedastic curve of Y on X. The beta-curves $y = \{\mu_3(x)\}^2/\{\mu_2(x)\}^3$ and $y = \mu_4(x)/\{\mu_2(x)\}^2$, using the measures of skewness and kurtosis, are the clitic curve and the kurtic curve, respectively.

Suppose that $Y = \tilde{y}_x$ is a solution of the equation

$$K(Y|x) = \tfrac{1}{2};$$

then $y = \tilde{y}_x$ is the median regression of Y on X. The scedastic curve of Y on X based on the mean deviation is

$$\mathrm{E}[\{|Y - \tilde{y}_x|\}|x].$$

The scedastic curve based on the semi-interquartile range is simply given by

$$y = \tfrac{1}{2}\{Q_3(x) - Q_1(x)\},$$

where $K\{Q_1(x)|x\} = \frac{1}{4}$ and $K\{Q_3(x)|x\} = \frac{3}{4}$.

We shall denote the normal distribution with mean μ and variance σ^2 by $N(\mu, \sigma^2)$. Further, the p.d.f. corresponding to $N(0,1)$ will be denoted by $b(x)$ and the d.f. by $B(x)$. We shall write the bivariate normal distribution with means μ and ν, variances σ_1^2 and σ_2^2, and correlation coefficient ρ as $N(\mu, \nu, \sigma_1^2, \sigma_2^2, \rho)$. When the margins are standardized variables, then $b(x, y; \rho)$ and $B(x, y; \rho)$ will denote the p.d.f. and the d.f., respectively.

CHAPTER 2

THE DIFFERENTIAL EQUATIONS METHOD

2.1 Pearson's equations

We know that the univariate Pearson differential equation

$$\frac{1}{f}\frac{df}{dx} = \frac{x - a}{b_0 + b_1 x + b_2 x^2} \tag{2.1.1}$$

has been useful in providing a great variety of frequency curves of practical utility, and extending this approach could prove to be fruitful. The form of this equation can be obtained as a limit of the slope–ordinate relationship of the univariate hypergeometric distribution. It is therefore natural to expect that a general family of surfaces might be developed from the form of the differential equations derived from the slope–ordinate ratios for the bivariate hypergeometric distribution.

Now, the probability function of the bivariate hypergeometric distribution is

$$z(x,y) = \binom{Np_1}{x}\binom{Np_2}{y}\binom{N - Np_1 - Np_2}{n - x - y}\Big/\binom{N}{n}. \tag{2.1.2}$$

We can take the partial difference coefficient of $z(x,y)$ with respect to x as

$$\begin{aligned}\Delta_x z(x,y) &= \tfrac{1}{2}\{z(x+1, y+\tfrac{1}{2}) - z(x, y+\tfrac{1}{2})\}\\ &= \tfrac{1}{4}\{z(x+1, y+1) + z(x+1, y) - z(x, y+1) - z(x,y)\}.\end{aligned} \tag{2.1.3}$$

Further,

$$z(x+\tfrac{1}{2}, y+\tfrac{1}{2}) = \tfrac{1}{4}\{z(x+1, y+1) + z(x+1, y) + z(x, y+1) + z(x,y)\}. \tag{2.1.4}$$

From (2.1.3) and (2.1.4), we have

$$\frac{\Delta_x z(x,y)}{z(x+\frac{1}{2}, y+\frac{1}{2})} = \frac{R(x,y) - 1}{R(x,y) + 1}, \tag{2.1.5}$$

where $R(x,y) = \dfrac{z(x+1, y+1) + z(x+1, y)}{z(x, y+1) + z(x,y)}$.

From (2.1.2) we find that

$$R(x,y) = \frac{(Np_1 - x)(n - x - y)\,Q(x,y)}{(x+1)(N - Np_1 - Np_2 - n + x + y + 2)\,Q(x-1, y)}, \tag{2.1.6}$$

where

$$Q(x,y) = (n-x-y-1)\,(Np_2-y) + (y+1)\,(N-Np_1-Np_2-n+x+y+2)$$

is a quadratic function in x and y. On substituting for $R\,(x,y)$ from (2.1.6) in (2.1.5), we see that

$$\frac{\Delta_x\, z\,(x,y)}{z\,(x+\frac{1}{2},\, y+\frac{1}{2})} = \frac{\text{cubic in } x,y}{\text{quartic in } x,y}\,. \tag{2.1.7}$$

Similarly, we find that

$$\frac{\Delta_y\, z\,(x,y)}{z\,(x+\frac{1}{2},\, y+\frac{1}{2})} = \frac{\text{another cubic in } x,y}{\text{same quartic in } x,y}\,. \tag{2.1.8}$$

Hence (2.1.7) and (2.1.8), in the limiting case, give the differential equations

$$\frac{1}{z}\frac{\partial z}{\partial x} = \frac{C_1}{QR}, \qquad \frac{1}{z}\frac{\partial z}{\partial y} = \frac{C_2}{QR}, \tag{2.1.9}$$

where C_1 and C_2 are two cubic functions in x and y, and QR is a quartic function in x and y.

It can be verified that the bivariate normal (§ 10.6), Plackett's uniform and Morgenstern's uniform (§ 10.7), the bivariate Cauchy (§ 10.8), Gumbel's exponential (§ 10.11), the Pareto (§ 10.12), Student's distribution (§ 10.13), the bivariate Type II distribution (§ 10.15) and Rhodes' distribution (§ 10.16) are solutions of these equations. It can further be shown that the distribution (Sagrista, 1952)

$$z\,(x,y) = \text{const. } e^{-Q}\,(h^2+Q)^n, \quad n > 0,$$

Q being a quadratic form, is also a solution. The margins of this distribution are complicated, and no application for general n is known.

From (2.1.9), we have

$$\frac{1}{k(y|x)}\,\frac{\partial k(y|x)}{\partial y} = \frac{C_2}{QR},$$

so that the conditional distributions include Pearson curves, but no conclusion can be drawn about the form of the margins.

The differential equations (2.1.9) and their motivation are due to Karl Pearson. The motivation was first stated in K. Pearson (1923a) and the equations were published in Rhodes (1923). Karl Pearson devoted himself to this problem as early as 1890 after publication of his univariate system, but the efforts to integrate these equations have not been rewarding. A historical account is given in K.Pearson (1923a).

2.2 Van Uven's equations

2.2a Definition

More recently, van Uven (1947 a,b; 1948 a,b) returned to the differential equation method. He started with the equations

$$\frac{1}{z}\frac{\partial z}{\partial x} = \frac{L_1}{Q_1}, \qquad \frac{1}{z}\frac{\partial z}{\partial y} = \frac{L_2}{Q_2}, \tag{2.2.1}$$

where L_1 and L_2 are linear functions of x and y, and Q_1 and Q_2 are quadratic functions of x and y. He did not link these equations with any hypergeometric series, and first we develop a motivation.

In § 2.1, we took the slope–ordinate ratios of the bivariate hypergeometric distribution at the point $(x+\frac{1}{2}, y+\frac{1}{2})$. In the univariate case, if we form the slope–ordinate ratio for the univariate hypergeometric distribution at x, we are again led to the univariate Pearson differential equation (e.g. Kendall and Stuart, 1969, p. 140) so that, instead of (2.1.5), we may take

$$\frac{\Delta_x z(x,y)}{z(x,y)} = \frac{z(x+1,y) - z(x,y)}{z(x,y)} .$$

From (2.1.2), we see that

$$\frac{\Delta_x z(x,y)}{z(x,y)} = \frac{(nNp_1 - N + n + Np_1 + Np_2 - 1) + x(Np_2 - N - 2) - y(Np_1 + 1)}{(x+1)(N - n - Np_1 - Np_2 + 1 + x + y)}$$

and $\dfrac{\Delta_y z(x,y)}{z(x,y)}$ can be written in a similar way. In the limiting case, these obviously lead to equations (2.2.1).

2.2b Solutions

Van Uven (1947b) has integrated these equations completely, and we proceed to give his results.

The functions L_1, L_2, Q_1 and Q_2 should satisfy the condition

$$\frac{\partial^2 \log z}{\partial x\,\partial y} = \frac{\partial}{\partial y}\left(\frac{L_1}{Q_1}\right) = \frac{\partial}{\partial x}\left(\frac{L_2}{Q_2}\right),$$

or
$$Q_2 \frac{\partial L_1}{\partial y} - Q_1 \frac{\partial L_2}{\partial x} = \frac{L_1 Q_2}{Q_1} \frac{\partial Q_1}{\partial y} - \frac{L_2 Q_1}{Q_2} \frac{\partial Q_2}{\partial x} . \quad (2.2.2)$$

Since, $\partial L_1/\partial y$ and $\partial L_2/\partial x$ are constants, the left-hand side of this expression is a quadratic function, and the requirement that the right-hand side must also be a quadratic function restricts the possibilities for L_1, L_2, Q_1 and Q_2.

Let us first consider the case when Q_1 and Q_2 have no common factors, so that Q_1/Q_2 and Q_2/Q_1 are irrational. Hence (2.2.2) can only be satisfied if

$$\frac{\partial Q_1}{\partial y} = 0, \qquad \frac{\partial Q_2}{\partial x} = 0.$$

Then Q_1 is a function of x only and Q_2 of y only, i.e.

$$Q_1 = Q_1(x), \quad Q_2 = Q_2(y). \quad (2.2.3)$$

On using these results in (2.2.2), we have

$$Q_2(y) \frac{\partial L_1}{\partial y} = Q_1(x) \frac{\partial L_2}{\partial x},$$

which is only possible if $\frac{\partial L_1}{\partial y} = \frac{\partial L_2}{\partial x} = 0$, so that

$$L_1 = L_1(x), \quad L_2 = L_2(y). \quad (2.2.4)$$

Substituting (2.2.3) and (2.2.4) in (2.2.1), we find that

$$\frac{\partial \log z}{\partial x} = \frac{L_1(x)}{Q_1(x)}, \qquad \frac{\partial \log z}{\partial y} = \frac{L_2(y)}{Q_2(y)},$$

which implies that X and Y are independently distributed. Hence, if Q_1 and Q_2 have no common factors, (2.2.2) implies that X and Y are independent.

Thus the nature of the solution of (2.2.1) depends mainly on the structure of Q_1 and Q_2, and the usefulness of the solution will depend on their having common factors. Important cases are

(I) Q_1 and Q_2 have one common factor.
(II) Q_1 and Q_2 are identical.
(III) Q_2 is a linear factor of Q_1.

Case I. Let $Q_1 = LL'$, $Q_2 = LL''$, so that (2.2.2) reduces to

$$L''\left(L\frac{\partial L_1}{\partial y} - L_1\frac{\partial L}{\partial y}\right) - L'\left(L\frac{\partial L_2}{\partial x} - L_2\frac{\partial L}{\partial x}\right) = \frac{LL''L_1}{L'}\frac{\partial L'}{\partial y} - \frac{LL'L_2}{L''}\frac{\partial L''}{\partial x} . \quad (2.2.5)$$

Again, since the left-hand side is a rational function, the right-hand side must be rational. Therefore, $\partial L'/\partial y = \partial L''/\partial x = 0$, so that $L' = L'(x)$ and $L'' = L''(y)$. On substituting these in (2.2.5), we get

$$L_1 = c_1 L + c_2 L'(x), \quad L_2 = d_1 L + d_2 L''(y),$$

where c_1, c_2, d_1, d_2 are constants. Hence

$$\frac{\partial \log z}{\partial x} = \frac{a_1}{L'(x)} + \frac{b_1}{L}, \qquad \frac{\partial \log z}{\partial y} = \frac{a_2}{L''(y)} + \frac{b_2}{L},$$

so that

$$\log z = p_1 \log L'(x) + p_2 \log L''(y) + p_3 \log L + \log k_0 ,$$

or $$z(x,y) = k_0 (ax+b)^{p_1} (cy+d)^{p_2} (a_1 x + b_1 y + c_1)^{p_3}. \tag{2.2.6}$$

This form contains the bivariate beta (§ 10.9), the Pareto (§ 10.12) and the F-distribution (§ 10.14).

Case II. Let $Q_1 = Q_2 = Q$, so that (2.2.2) becomes

$$cQ = L_1 \frac{\partial Q}{\partial y} - L_2 \frac{\partial Q}{\partial x}, \tag{2.2.7}$$

where c is a constant. If $c \neq 0$ then the conic $Q = 0$ passes through the point of intersection of $\partial Q/\partial x = \partial Q/\partial y = 0$ and so through its centre. Hence Q degenerates into a pair of straight lines, which is contrary to our assumption. Hence we have $c = 0$, so that from (2.2.7)

$$L_1 = n\frac{\partial Q}{\partial x}, \quad L_2 = n\frac{\partial Q}{\partial y},$$

which gives $\log z = \log k_0 + n \log Q$, or

$$z(x,y) = k_0 \{ax^2 + 2hxy + by^2 + 2gx + 2fy + c\}^n. \tag{2.2.8}$$

This contains the bivariate Cauchy (§ 10.8), Student's distribution (§ 10.13), and the Type II distribution (§ 10.15).

Case III. We have $Q_1 = LL'$, $Q_2 = L$. On following the argument used in Case I, we get

$$z(x,y) = k_0 (ax+b)^p (a_1 x + b_1 y + c_1)^q e^{-cy}. \tag{2.2.9}$$

This contains McKay's gamma distribution (§ 10.10).

In fact, van Uven has considered all possible cases, but other solutions do not lead to distributions with both margins of the same form. Moreover, the method is similar to that used in Case I.

2.2c Properties

We now proceed to investigate general properties of the family. From (2.2.1), we have

$$\frac{1}{k(y|x)}\frac{\partial k(y|x)}{\partial y} = \frac{L_2}{Q_2}, \tag{2.2.10}$$

so that the conditional distributions belong to the Pearson family. Let $\{a(x),\ b(x)\}$ be the range of y for $k(y|x)$. From (2.2.10), we have immediately,

$$\int_{a(x)}^{b(x)} Q_2 \frac{\partial k(y|x)}{\partial y} dy = \int_{a(x)}^{b(x)} L_2\, k(y|x) dy.$$

Integrating the left-hand side by parts, we find, under the assumption

$$\lim_{y\to a(x)} y^2 k(y|x) = \lim_{y\to b(x)} y^2 k(y|x) = 0, \tag{2.2.11}$$

that

$$-\int_{a(x)}^{b(x)} \frac{\partial Q_2}{\partial y} k(y|x) dy = \int_{a(x)}^{b(x)} L_2 k(y|x) dy.$$

Hence $E(Y|x)$ is of the form $ax + b$, so that the regression of Y on X is linear. Multiplying (2.2.10) by y and proceeding along the same lines, we see, under an assumption analogous to (2.2.11), that $\operatorname{var}(Y|x)$ is parabolic. Similarly, the regression of X on Y is linear and $\operatorname{var}(X|y)$ is parabolic. Further, let

$$\lim_{y\to a(x)} y^{s+2} k(y|x) = \lim_{y\to b(x)} y^{s+2} k(y|x) = 0$$

and assume that similar conditions hold for $x^{r+2}k(x|y)$. We then have

$$E(X^r Y^s L_1') = -rE(X^{r-1}Y^s Q_1),\ E(X^r Y^s L_2') = -sE(X^r Y^{s-1} Q_2),$$

where
$$L_1' = L_1 + \frac{\partial Q_1}{\partial x},\quad L_2' = L_2 + \frac{\partial Q_2}{\partial y}.$$

Hence the constants in the differential equations (2.2.1) can be expressed in terms of moments, but this study has not been pursued. Alternatively, we could first fit the margins and then proceed with fitting the bivariate distribution.

It can be seen that the distribution is unimodal. The mode being the solution of $L_1 = L_2 = 0$. Van Uven (1948a) has obtained the saddle-point curve,

$$\left(\frac{\partial^2 z}{\partial x^2}\right)\left(\frac{\partial^2 z}{\partial y^2}\right) = \left(\frac{\partial^2 z}{\partial x\, \partial y}\right)^2$$

for Case I and Case II. The curves turn out to be ellipses as in the bivariate normal case.

2.3 Remarks

We thus find that Pearson's equations and van Uven's equations contain important distributions. A more elastic system is naturally represented by

$$\frac{1}{z}\frac{\partial z}{\partial x} = \frac{C_1}{QR_1}, \qquad \frac{1}{z}\frac{\partial z}{\partial y} = \frac{C_2}{QR_2},$$

where QR_1 and QR_2 are two quartics in x and y. This will include solutions of both systems, but fitting will be more difficult since the number of parameters has increased.

CHAPTER 3

BIVARIATE EDGEWORTH EXPANSION

3.1 The expansion

A well-known expansion in the univariate case is the Gram–Charlier and Edgeworth expansion, and we discuss here an extension. The work was started as early as 1896 by Edgeworth and he published further papers in 1905 and 1917. Van der Stok (1908), Charlier (1914), Jørgensen (1916), Wicksell (1917 a,b), K. Pearson (1925), and others also contributed to the field in its developing stages. Pretorius (1930) gives an account of these developments.

Let

$$b(x,y;\rho) = \frac{1}{2\pi\sqrt{(1-\rho^2)}}\exp\left\{-\frac{1}{2(1-\rho^2)}(x^2-2\rho xy+y^2)\right\},$$

and let $h(x,y)$ be the p.d.f. of the r.v. (X,Y) with $\mu=\nu=0$, $\sigma_1=\sigma_2=1$. Following Kendall (1949), we show that $h(x,y)$ can be expanded in the following series of derivatives of $b(x,y;\rho)$.

$$h(x,y) = b(x,y;\rho) + \sum_{r+s\geqslant 3}(-1)^{r+s}A_{rs}\frac{D_1^r}{r!}\frac{D_2^s}{s!}b(x,y;\rho), \quad (3.1.1)$$

where $D_1=\partial/\partial x$, $D_2=\partial/\partial y$ and A_{rs} are functions of the cumulants of X and Y.

Let κ_{rs} be the cumulant of order (r,s) of the r.v. (X,Y). We first show that

$$h(x,y) = \exp\left\{\kappa_{11}D_1D_2 + \sum_{r+s\geqslant 3}(-1)^{r+s}\kappa_{rs}\frac{D_1^r}{r!}\frac{D_2^s}{s!}\right\}b(x)b(y), \quad (3.1.2)$$

provided the series is convergent and the distribution is uniquely determined by its moments. We proceed along the same lines as in the univariate case (Kendall and Stuart, 1969, pp. 157–8). The characteristic function of $\exp\{\kappa_{rs}D_1^rD_2^s b(x)b(y)\}$ is

$$\int_{-\infty}^{\infty}\int_{-\infty}^{\infty}e^{it_1x+it_2y}\exp\{\kappa_{rs}D_1^rD_2^s b(x)b(y)\}\,dx\,dy$$

$$= \sum_{j=0}^{\infty}\frac{\kappa_{rs}^j}{j!}\left\{\int_{-\infty}^{\infty}e^{it_1x}D_1^{rj}b(x)dx\right\}\left\{\int_{-\infty}^{\infty}e^{it_2y}D_2^{rj}b(y)dy\right\}$$

$$= \sum_{j=0}^{\infty} \frac{\kappa_{rs}^j}{j!} \{\sqrt{(2\pi)}(-i)^{rj} t_1^{rj} b(t_1)\} \times \{\sqrt{(2\pi)}(-i)^{sj} t_2^{sj} b(t_2)\}$$

$$= 2\pi b(t_1) b(t_2) \exp\{\kappa_{rs}(-i)^{r+s} t_1^r t_2^s\}. \quad (3.1.3)$$

Hence the characteristic function of (3.1.2) is given by

$$\exp\left\{-\tfrac{1}{2}(t_1^2 + 2\rho t_1 t_2 + t_2^2) + \sum_{r+s\geqslant 3} \frac{\kappa_{rs}}{r!s!} (it_1)^r (it_2)^s\right\}.$$

Thus the κ_{rs}'s are in fact the cumulants of the distribution and (3.1.2) holds under the conditions stated.

Further, from (3.1.3), the characteristic function of $\exp\{\kappa_{11} D_1 D_2\} b(x)b(y)$, κ_{11} being ρ, is

$$\exp\{-\tfrac{1}{2}(t_1^2 + 2\rho t_1 t_2 + t_2^2)\}$$

so that

$$\exp\{\kappa_{11} D_1 D_2\} b(x)b(y) = b(x,y;\rho)$$

and consequently (3.1.2) can be written as

$$h(x,y) = \exp\left\{\sum_{r+s\geqslant 3} (-1)^{r+s} \kappa_{rs} \frac{D_1^r}{r!} \frac{D_2^s}{s!}\right\} b(x,y;\rho). \quad (3.1.4)$$

Finally, on expanding (3.1.4), we obtain (3.1.1).

We now verify that the margins of (3.1.1) are Gram–Charlier and Edgeworth series. Observing that

$$\int_{-\infty}^{\infty} D_1^r D_2^s\, b(x,y;\rho)\,dy = D_1^r b(x), \quad \text{if } s = 0,$$

$$= [D_1^r D_2^{s-1} b(x,y;\rho)]_{y=-\infty}^{y=\infty} = 0 \quad \text{if } s \geqslant 1,$$

we see from (3.1.1) that the marginal p.d.f. of X is

$$f(x) = \left[1 + \sum_{r=3}^{\infty} \frac{A_{r0}}{r!} H_r(x)\right] b(x), \quad (3.1.5)$$

where $H_r(x)$ is the Hermite polynomial of the rth order and

$$D_1^r b(x) = (-1)^r H_r(x)\, b(x). \quad (3.1.6)$$

Similarly, the joint distribution function is

$$H(x,y) = B(x,y;\rho) + \sum_{r+s\geqslant 3} (-1)^{r+s} \frac{A_{rs}}{r!s!} D_1^{r-1} D_2^{s-1} b(x,y;\rho).$$

It should be noted that the differential equations method lacks this simplicity.

Good (1957, 1961) has extended the saddle-point method of Daniels (1954), which is a generalization of the Edgeworth series approach; but applications to fitting bivariate distributions have not been attempted. Chambers (1967) has given numerical methods of approximating multivariate distributions with known moments by multivariate Edgeworth expansions.

3.2 Coefficients

To see the order of magnitude of the terms in the Edgeworth expansion (3.1.1), it is reasonable to assume, as in the univariate case, that the standardized cumulants κ_{rs} are of the form $l_{rs}/n^{[(r+s)/2]-1}$. Substituting for κ_{rs} in (3.1.4) and comparing the coefficients of (3.1.1) and (3.1.4), we find that the terms in (3.1.1) for $r+s=3$ are of the order $n^{-1/2}$, for $r+s=4,6$ are of the order n^{-1} and so on. In fact,

$$\begin{aligned}
\text{for } r+s=3:\ & A_{30} = \kappa_{30} = \mu_{30}, \quad A_{21} = \kappa_{21} = \mu_{21}. \\
\text{for } r+s=4:\ & A_{40} = \kappa_{40} = \mu_{40} - 3, \quad A_{31} = \kappa_{31} = \mu_{31} - 3\rho, \\
& A_{22} = \kappa_{22} = \mu_{22} - 2\rho^2 - 1. \\
\text{for } r+s=6:\ & A_{60} = 10\kappa_{30}^2, \quad A_{51} = 10\kappa_{30}\kappa_{21}, \\
& A_{42} = 4\kappa_{30}\kappa_{12} + 6\kappa_{21}^2, \\
& A_{33} = \kappa_{30}\kappa_{03} + 9\kappa_{12}\kappa_{21}.
\end{aligned} \tag{3.2.1}$$

The distribution given by considering terms up to $r+s=3$ was first investigated by Edgeworth (1896), and Wicksell (1917a) added the terms for $r+s=4,6$. For terms up to $r+s=4$, K. Pearson (1925) obtained the distribution by approximating a generalized bivariate hypergeometric model and called it a "fifteen constant" bivariate distribution; the fifteen constants being the constant of total frequency, the first four moments of each margin and $\kappa_{11}, \kappa_{12}, \kappa_{21}, \kappa_{22}, \kappa_{31}, \kappa_{13}$. We shall call it a Type AA surface. The fitting of this distribution can best be performed after the p.d.f. has been expanded as a polynomial in x and y; that is,

$$h(x,y) = b(x,y;\rho)\ \{1 - a_0 + a_1\, x + a_2\, y + b_1\, x^2 + 2b_2\, xy + b_3\, y^2$$
$$+ c_1\, x^3 + c_2\, x^2y + c_3\, xy^2 + c_4\, y^3$$
$$+ d_1\, x^4 + d_2\, x^3y + 3d_3\, x^2y^2 + d_4\, xy^3 + d_5\, y^4\}. \qquad (3.2.2)$$

Let $a = (1-\rho^2)^{-1}$, $B_{20} = \frac{1}{24}(\beta_{20}-3)$, $B_{02} = \frac{1}{24}(\beta_{02}-3)$,

$$Q_{31} = \tfrac{1}{6}(\mu_{31}-3\rho), \quad Q_{13} = \tfrac{1}{6}(\mu_{13}-3\rho), \quad Q_{22} = \tfrac{1}{12}(\mu_{22}-1-2\rho^2). \qquad (3.2.3)$$

We then have

$$a_0 = -3a^2\{B_{20} + B_{02} - \rho(Q_{31} + Q_{13}) + (1+2\rho^2)Q_{22}\},$$
$$a_1 = \tfrac{1}{2}a^2(3\rho\,\mu_{21} - \sqrt{\beta_{10}} - (1+2\rho^2)\mu_{12} + \rho\sqrt{\beta_{01}}\},$$
$$b_1 = -3a^3\{2B_{20} + 2\rho^2 B_{02} - 2\rho Q_{31} - \rho(1+\rho^2)Q_{13} + (1+5\rho^2)Q_{22}\},$$
$$b_2 = \tfrac{3}{2}a^3\{4\rho(B_{20} + B_{02}) - (1+3\rho^2)(Q_{31} + Q_{13}) + 4\rho(2+\rho^2)Q_{22}\},$$
$$c_1 = \tfrac{1}{6}a^3\{\rho^2(3\mu_{12} - \rho\sqrt{\beta_{01}}) - (3\rho\,\mu_{21} - \sqrt{\beta_{10}})\},$$
$$c_2 = \tfrac{1}{2}a^3\{(1+2\rho^2)\mu_{21} - \rho\sqrt{\beta_{10}} - \rho(2+\rho^2)\mu_{12} + \rho^2\sqrt{\beta_{01}}\},$$
$$d_1 = a^4(B_{20} + \rho^4 B_{02} - \rho Q_{31} - \rho^3 Q_{13} + 3\rho^2 Q_{22}),$$
$$d_2 = -a^4\{4\rho B_{20} + 4\rho^3 B_{02} - (1+3\rho^2)Q_{31} - \rho^2(3+\rho^2)Q_{13}$$
$$+ 6\rho(1+\rho^2)Q_{22}\},$$
$$d_3 = a^4\{2\rho^2(B_{20} + B_{02}) - \rho(1+\rho^2)(Q_{31} + Q_{13}) + (1+4\rho^2+\rho^4)Q_{22}\},$$

and other constants can be obtained by interchanging the suffixes.

For the Type AA surface, Wicksell (1917b) gives an approximate solution to the problem of the curves of equal probability. For moderately skew distributions, he finds that in the neighbourhood of the mode the curves of equal probabilities are ellipses, while further away they are disturbed ellipses. He provides the necessary tables to construct these outer ellipses.

3.3 The conditional moments

We first obtain $E[H_k(aY)\,|\,x]$ where now $a = (1-\rho)^{-1/2}$ and $H_k(.)$ is the Hermite polynomial of the kth order defined at (3.1.6). We will then derive particular moment curves.

From (3.1.1) and (3.1.5), we have

$$E[H_k(aY)\,|\,x]f(x) = I_1 + \sum_{r+s\geqslant 3} (-1)^{r+s}\frac{A_{rs}}{r!s!}I_{rs}, \qquad (3.3.1)$$

where $$I_1 = \int_{-\infty}^{\infty} H_k(ay)\, b(x,y;\rho)\, dy \tag{3.3.2}$$

and $$I_{rs} = \int_{-\infty}^{\infty} H_k(ay)\, D_1^r\, D_2^s\, b(x,y;\rho)\, dy. \tag{3.3.3}$$

Using

$$b(x,y;\rho) = ab(x)b\{a(y-\rho x)\}$$

in (3.3.3) and making the transformation $y' = a(y-\rho x)$, we find that

$$I_{rs} = a^s\, D_1^r \left\{ b(x) \int_{-\infty}^{\infty} H_k(y + a\rho x)\, D_2^s\, b(y)\, dy \right\}. \tag{3.3.4}$$

We now show that

$$H_k(y + a\rho x) = \sum_{i=0}^{k} \binom{k}{i} (a\rho x)^i\, H_{k-i}(y). \tag{3.3.5}$$

Let $c = a\rho x$. We have from the generating function of $H_k(y+c)$,

$$\sum_{k=0}^{\infty} \frac{H_k(y+c)}{k!} t^k = e^{-(t^2/2)+ty+tc}$$

$$= e^{-(t^2/2)+ty} e^{tc}$$

$$= \sum_{i=0}^{\infty} \frac{H_i(y) t^i}{i!} \sum_{j=0}^{\infty} \frac{t^j c^j}{j!}.$$

On equating coefficients of t^k on both sides, we obtain (3.3.5). Using (3.3.5) in (3.3.4), we find that

$$I_{rs} = (-1)^s\, a^s\, D_1^r \left\{ b(x) \sum_{i=0}^{k} \binom{k}{i} (a\rho x)^i \int_{-\infty}^{\infty} H_{k-i}(y) H_s(y) b(y)\, dy \right\}.$$

In virtue of the orthogonal property of $H_k(y)$, it follows that

$$\left.\begin{aligned} I_{rs} &= (-1)^s\, a^k\, \rho^{k-s}\{k!/(k-s)!\}\, D_1^r\{b(x)x^{k-s}\}, \quad \text{if } s \leqslant k, \\ &= 0, \text{ if } s > k. \end{aligned}\right\} \tag{3.3.6}$$

From the identity,

$$e^{\frac{1}{2}t^2} \sum_{r=0}^{\infty} \frac{H_r(x) t^r}{r!} = e^{tx},$$

we have on equating coefficients of t^k on both sides

$$x^k = \sum_{m=0}^{[\frac{1}{2}k]} \frac{k!\,2^{-m}}{m!\,k-2m!} H_{k-2m}(x).$$

Using this expansion in (3.3.6), we find that

$$\begin{aligned} I_{rs} &= b(x)\,(-1)^{r+s}\,\rho^{k-s}\,a^k\,k! \sum_{m=0}^{[(k-s)/2]} \frac{2^{-m} H_{k+r-s-2m}(x)}{m!\,k-s-2m!}, \quad s \leqslant k, \\ &= 0,\ s > k. \end{aligned} \tag{3.3.7}$$

Now $I_1 = I_{00}$, so that from (3.3.6),

$$I_1 = b(x)\,(a\rho x)^k. \tag{3.3.8}$$

Finally, on substituting (3.3.7) and (3.3.8) in (3.3.1), we obtain

$$\mathrm{E}[H_k(aY)\,|\,x]$$

$$= \frac{a^k}{D(x)} \left\{ (\rho x)^k + k! \sum_{\substack{r+s \geqslant 3 \\ s \leqslant k}} \sum_{m=0}^{[(k-s)/2]} \frac{2^{-m} A'_{rs}\,\rho^{k-s} H_{k+r-s-2m}(x)}{m!\,k-s-2m!} \right\} \tag{3.3.9}$$

where $A'_{rs} = A_{rs}/(r!\,s!)$ and $D(x) = 1 + \sum_{r=3}^{\infty} A'_{r0} H_r(x)$.

For $k = 1$, we have from (3.3.9), after using $H_1(y) = y$, that

$$\mathrm{E}(Y|x) = \left\{ \rho x + \sum_{r=3}^{\infty} \left(\rho A'_{r0} H_{r+1}(x) + A'_{r-1,1} H_{r-1}(x) \right) \right\} \Big/ D(x). \tag{3.3.10}$$

In particular, for the Type AA surface this becomes

$$\mathrm{E}(Y\,|\,x) = \rho x + \frac{B_x}{A_x} \tag{3.3.11}$$

where

$$\begin{aligned} A_x &= 1 + \tfrac{1}{6}\sqrt{\beta_{10}}\,H_3(x) + \tfrac{1}{24}(\beta_{20} - 3)\,H_4(x) \\ &\equiv 1 + a_3 H_3(x) + a_4 H_4(x), \end{aligned}$$

and

$$\begin{aligned} B_x &= \tfrac{1}{2}(\mu_{21} - \rho\sqrt{\beta_{10}})\,H_2(x) + \tfrac{1}{6}(\mu_{31} - \rho\beta_{20})\,H_3(x) \\ &\equiv b_2 H_2(x) + b_3 H_3(x). \end{aligned}$$

Further, on putting $k = 2$ in (3.3.9) and using $H_2(y) = y^2 - 1$ we find for the Type AA surface

$$\mathrm{var}(Y \mid x) = 1 - \rho^2 - \frac{C_x}{A_x} - \left(\frac{B_x}{A_x}\right)^2, \tag{3.3.12}$$

where

$$\begin{aligned} C_x &= \{\rho(\mu_{21} - \rho\sqrt{\beta_{10}}) - (\mu_{12} - \rho\mu_{21})\} H_1(x) \\ &\quad + \tfrac{1}{2}\{2\rho(\mu_{31} - \rho\beta_{20}) - \mu_{22} + 1 + \rho^2(\beta_{20} - 1)\} H_2(x) \\ &\equiv c_1 H_1(x) + c_2 H_2(x). \end{aligned}$$

For the Type AA surface, we also have

$$\mu_3(y) = \frac{D_x}{A_x} + 3\left(\frac{B_x}{A_x}\right)\left(\frac{C_x}{A_x}\right) + 2\left(\frac{B_x}{A_x}\right)^3, \tag{3.3.13}$$

where

$$\begin{aligned} D_x &= \{\sqrt{\beta_{01}} - \rho^3\sqrt{\beta_{10}} - 3\rho(\mu_{12} - \rho\mu_{21})\} \\ &\quad + 6\{Q_{13} - 6\rho Q_{22} + 3\rho^2 Q_{31} - 4\rho^3 B_{20}\} H_1(x) \\ &\equiv d_0 + d_1 H_1(x) \end{aligned}$$

and Q_{13}, etc. are given by (3.2.3), and

Table 3.1 Frequencies of the number of bean
Johanssen's data,

Y ↓	X→ 17	16·5	16	15·5	15	14·5	14	13·5
9·125		2			3			
8·875	4	8	17	19				
8·625	2	23	101	156	93	23	2	
8·375		18	105	494	574	227	56	9
8·125		4	44	375	956	913	362	7[illegible]
7·875			7	81	385	871	794	33[illegible]
7·625			1	4	65	236	469	36[illegible]
7·375					6	23	91	13[illegible]
7·125						1	13	1[illegible]
6·875								[illegible]
6·625								
6·375								
Totals	6	55	275	1129	2082	2294	1787	929

$$\mu_4(y) = 3(1-\rho^2)^2 + \frac{e_0}{A_x} - 6(1-\rho^2)\left\{\frac{C_x}{A_x} + \left(\frac{B_x}{A_x}\right)^2\right\}$$
$$- 4\left(\frac{B_x}{A_x}\right)\left(\frac{D_x}{A_x}\right) - 6\left(\frac{B_x}{A_x}\right)^2\left(\frac{C_x}{A_x}\right) - 3\left(\frac{B_x}{A_x}\right)^4, \qquad (3.3.14)$$

where

$$e_0 = (\beta_{02} - 3)(1 - 4\rho^2) - 3\rho^4(\beta_{20} - 3) + 6\rho^2(\mu_{22} - 1 - 2\rho^2)$$
$$- 4\rho(\mu_{13} - \rho\beta_{02}) - 4\rho^3(\mu_{31} - \rho\beta_{20}).$$

The above forms of the conditional moment curves represent departures of the Type AA distribution from the normal distribution. If we take $\kappa_{rs} = l_{rs}/n^{[(r+s)/2]-1}$, in (3.3.11)–(3.3.14) we find on considering terms up to order $n^{-1/2}$ that $E(Y|x)$ is cubic, $\text{var}(Y|x)$ is parabolic, the skewness is linear, and the kurtosis is constant. Wicksell (1917a) obtained the first four conditional moments given by (3.3.11)–(3.3.14) and also investigated the form of these curves up to terms of order $n^{-1/2}$. The above form of (3.3.11) and (3.3.12) was obtained by K. Pearson (1925) and of (3.3.13) and (3.3.14) by Pretorius (1930).

of length X and breadth Y measured in millimetres cited by Pretorius (1930)

13	12·5	12	11·5	11	10·5	10	9·5	Totals
								5
								48
								400
								1483
12	3							2742
89	19	3						2579
175	55	27	4					1397
124	78	37	22	11		1		530
28	35	25	32	11	6	1		170
9	8	21	12	13	7	1		72
		2		1	4	3		10
	1				1	1	1	4
437	199	115	70	36	18	7	1	9440

3.4 Applications

3.4a Fitting

The Type AA distribution has been fitted by Wicksell (1917a) and K. Pearson (1925). Later, Pretorius (1930) critically examined the adequacy of the Type AA distribution by using five observed distributions ranging from moderately skew to considerably skew. He did not attempt to carry out the lengthy process of calculating cell frequencies, but he obtained the regression and scedastic curves in all five cases and the curves of skewness and kurtosis in two cases. Judged by these curves, the fit provided by Type AA distributions was reasonably good in some cases, but in others the fit was unsatisfactory and, in particular, the curves suffered from arbitrary oscillation. We reproduce an example.

Example 3.1. Table 3.1 shows the distribution of 9440 beans according to length (X) and breadth (Y). After using Sheppard's corrections, it is found that the moments are —

$$\mu = 14{\cdot}404{,}608; \quad \nu = 7{\cdot}975{,}530; \quad \rho = 0{\cdot}781{,}142;$$
$$\sigma_1 = 0{\cdot}899{,}781; \quad \sigma_2 = 0{\cdot}339{,}870; \quad q_{21} = -0{\cdot}653{,}267;$$
$$\sqrt{\beta_{10}} = -0{\cdot}910{,}569; \quad \sqrt{\beta_{01}} = -0{\cdot}440{,}832; \quad q_{12} = -0{\cdot}502{,}495;$$
$$\beta_{20} = 4{\cdot}862{,}944; \quad \beta_{02} = 3{\cdot}654{,}374; \quad q_{22} = 3{\cdot}167{,}011;$$
$$q_{31} = 3{\cdot}641{,}829; \quad q_{13} = 3{\cdot}072{,}626;$$

where $q_{rs} = \mu_{rs}/\sigma_1^r\,\sigma_2^s$.

On using these moments, we find that the coefficients of the Hermite polynomial terms in the expressions for the regression and scedastic curves, (3.3.11) and (3.3.12) respectively, are —

for Y on X: $a_3 = -0{\cdot}151{,}761$; $a_4 = 0{\cdot}077{,}623$; $b_2 = 0{\cdot}029{,}008$;
$b_3 = -0{\cdot}024{,}835$; $c_1 = 0{\cdot}037{,}520$; $c_2 = -0{\cdot}021{,}352$;

for X on Y: $a_3 = -0{\cdot}073{,}472$; $a_4 = 0{\cdot}027{,}256$; $b_2 = -0{\cdot}079{,}072$;
$b_3 = 0{\cdot}036{,}430$; $c_1 = 0{\cdot}137{,}215$; $c_2 = -0{\cdot}103{,}359$.

The regression curves and scedastic curves are shown in Fig. 3.1 to 3.4 and the plotted circles show the corresponding values calculated from the observed distribution. From Fig. 3.1 and 3.3 we observe that the regression curves are a good fit, except that the regression curve of length on breadth suffers from arbitrary oscillation for points in one of the tails. From Fig. 3.2 and 3.4 we see that both scedastic curves are oscillatory to a much higher degree and that the fit is unsatisfactory.

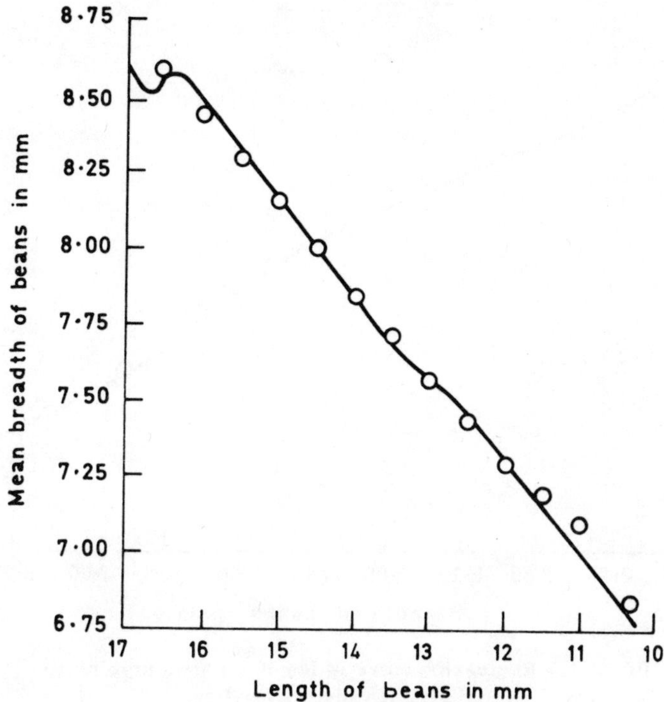

Fig. 3.1 Regression curve of breadth on length of beans (Type AA distribution)

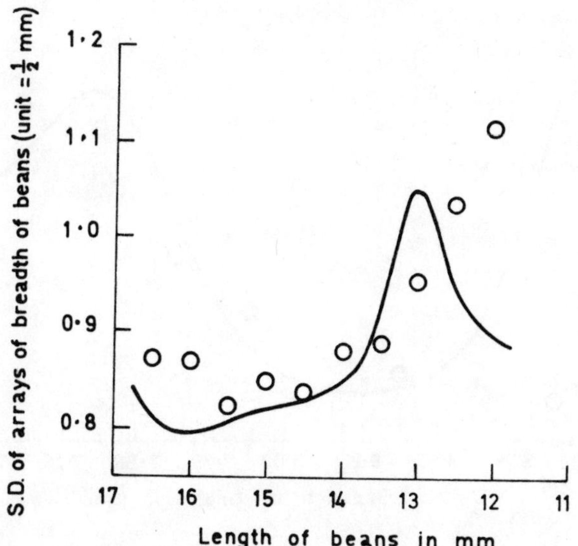

Fig. 3.2 Scedastic curve of breadth on length of beans (Type AA distribution)

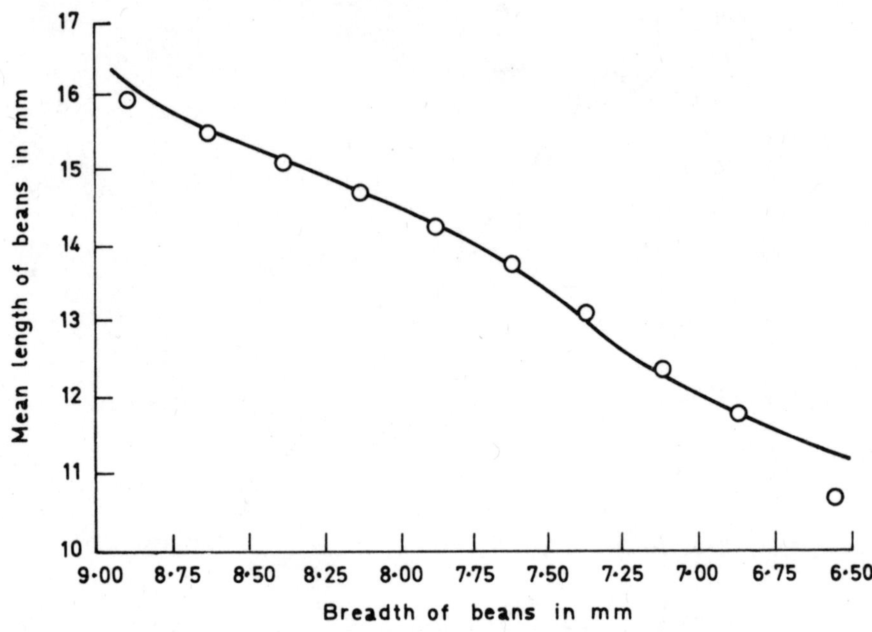

Fig. 3.3 Regression curve of length on breadth of beans (Type AA distribution)

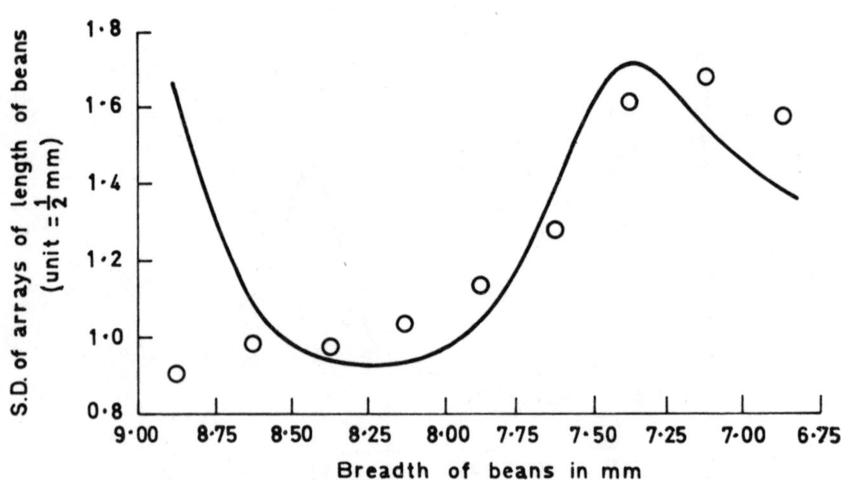

Fig. 3.4 Scedastic curve of length on breadth of beans (Type AA distribution)

We note that the marginal distribution of the length fitted by the Edgeworth series with terms up to $i = 5$ gives negative frequencies in the tail (Kendall and Stuart, 1969, p. 160), which implies that the Edgeworth surface for terms up to $i + j = 5$ will give some negative cell-frequencies in the corresponding tail-region.

3.4b Other applications

The distribution has also been used in the study of robustness of the usual normal theory tests. Kendall (1949) examined the effect of non-normality on Kendall's rank correlation coefficient t (the sample value of Kendall's τ) and Spearman's rank correlation coefficient, starting from the characteristic function and curtailing the series suitably. Gayen (1951) studied the robustness of the normal theory distribution of the ordinary correlation coefficient by considering samples from bivariate Edgeworth series with $r + s = 3, 4, 6$. He also studied the effect of non-normality on Fisher's z-transformation. Starting from the same expansion, Srivastava (1960) has studied the robustness of the normal theory distribution of the regression coefficient. There have been various other studies in the past, and references to the earlier work can be found in these papers.

3.5 Remarks

As in the univariate case, the sum of a finite number of terms of the expansion may give negative frequencies, particularly near the tails, as shown in Example 3.1; and truncating the series at $r + s = k$ may give a worse fit than truncating at $r + s = k - 1$. Besides, when we depart from the normal distribution in certain directions, the distribution can have more than one mode, or at least local maxima. There has been no investigation along the lines of the univariate study by Barton and Dennis (1952) to determine the range of values of the moments for which it is safe to use the bivariate Edgeworth expansion. Thus, not knowing the characteristics of the surface in the tail region, the range of validity of the conclusions of the robustness studies cannot be asserted. In addition, the method of moments, when used for fitting a bivariate Edgeworth expansion, can introduce arbitrary behaviour into the conditional moment-curves. Therefore, at present, this approach is unsatisfactory in either case.

CHAPTER 4

THE TRANSLATION METHOD

4.1 Introduction

We now consider the idea of transforming from a random variable (X, Y) to another random variable (X_0, Y_0) which has a more convenient distribution. For obvious reasons, we assume that (X_0, Y_0) is $N(0, 0, 1, 1, \rho)$. Edgeworth (1896) described this method as the translation method. Later on, Edgeworth (1914, 1917) took

$$x = a_0(x_0 + k_1 x_0^2 + \lambda_1 x_0^3 + \nu_1 y_0^2), \quad y = a_1(y_0 + k_2 y_0^2 + \lambda_2 y_0^3 + \nu_2 x_0^2).$$

In the case of the simple translation ($\nu_1 = \nu_2 = 0$), the moments are seen to satisfy the relations

$$\mu_{21} = \frac{\rho}{3}(2\mu_{30} + \rho\mu_{03}), \quad \mu_{12} = \frac{\rho}{3}(2\mu_{03} + \rho\mu_{30}).$$

By considering various observed distributions, Edgeworth (1917) and Pretorius (1930) found that these relations are too restrictive, and they concluded that the surface is inadequate for fitting purposes. For the composite translation ($\nu_1, \nu_2 \neq 0$), the estimation of the parameters by the method of moments is very tedious. We now proceed to give a more successful approach which was developed by Johnson (1949b).

4.2 Johnson's transformations

4.2a Definition

Johnson (1949a) proposed the frequency curves based on transformations of the form

$$x_0 = \gamma + \delta T\left(\frac{x' - \mu}{\lambda}\right) = \gamma + \delta T(x),$$

where X_0 is $N(0, 1)$, $T(.)$ is a monotone function, γ, δ, μ and λ are parameters, and X is a standardized random variable. The following special systems:

(1) the S_L (log-normal) system: $T_L(x) = \log x$,
(2) the S_B system: $T_B(x) = \log\{x/(1-x)\}$,
(3) the S_U system: $T_U(x) = \sinh^{-1} x$,

together with the original normal distribution S_N with $T_N(x) = x$, cover the entire (β_1, β_2) region.

Let (X_0, Y_0) be $N(0,0,1,1,\rho)$ with p.d.f. $b(x_0, y_0; \rho)$ and define the standardized random variable (X, Y) by the transformation

$$x_0 = \gamma_1 + \delta_1 T_I(x), \quad y_0 = \gamma_2 + \delta_2 T_J(y), \quad \delta_1, \delta_2 > 0, \tag{4.2.1}$$

where $I, J = N, L, B$ or U, and $\gamma_1, \gamma_2, \delta_1$ and δ_2 are constants. The p.d.f. (X, Y) is given by

$$h_{I,J}(x,y) = \delta_1 \delta_2 b(\gamma_1 + \delta_1 T_I(x), \gamma_2 + \delta_2 T_J(y); \rho) T_I'(x) T_J'(y). \tag{4.2.2}$$

We shall describe this system as an S_{IJ} system. We thus have ten different bivariate systems. S_{NN} is the bivariate normal distribution, S_{LL} is the bivariate log-normal distribution studied by Wicksell (1917c), and S_{NL} is the semi-logarithmic distribution studied by Yuan (1933).

4.2b The conditional distribution

Let us denote the p.d.f. of Y in the S_{IJ} system by $g_J(y; \gamma_2, \delta_2)$; then we show that the conditional p.d.f. of Y given X is simply

$$k_J(y|x) = g_J(y; \gamma', \delta'), \tag{4.2.3}$$

where

$$\gamma' = [\gamma_2 - \rho\{\gamma_1 + \delta_1 T_I(x)\}]/\sqrt{(1-\rho^2)} \tag{4.2.4}$$

and

$$\delta' = \delta_2/\sqrt{(1-\rho^2)}. \tag{4.2.5}$$

We know that the conditional distribution of Y_0 given X_0 is $N[\rho x_0, 1-\rho^2]$ so that for given X the random variable $\gamma_2 + \delta_2 T_J(Y)$ is $N[\rho\{\gamma_1 + \delta_1 T_I(x)\}, 1-\rho^2]$. Thus $\delta' + \gamma' T_J(Y)$ is $N(0,1)$ with δ' and γ' defined by (4.2.4) and (4.2.5). Consequently we obtain (4.2.3).

This result implies that the conditional distribution of Y given X is of the same type as the marginal distribution of Y. Since $T_I(x)$ varies from $-\infty$ to ∞ monotonically, γ' is a monotonic function of x and if $\rho > 0$, γ' is a decreasing function of x. Furthermore, γ' has a transition point at $x_0 = \gamma_2/\rho$. The parameter δ' is constant. We now examine the conditional moment curves of these systems.

4.2c The conditional moment curves

Case I. Let the marginal p.d.f. of Y be log-normal. From univariate moments and (4.2.3), we have

$$E(Y|x) = \omega\omega', \quad \text{var}(Y|x) = (\omega\omega')^2(\omega^2 - 1), \tag{4.2.6}$$

$$\beta_1(x) = (\omega^2 - 1)(\omega^2 + 2)^2,$$

$$\beta_2(x) = 3 + (\omega^2 - 1)(\omega^6 + 3\omega^4 + 6\omega^2 + 6), \tag{4.2.7}$$

where $\omega = \exp(1/2\delta'^2)$, $\omega' = \exp(-\nu'/\delta')$, (4.2.8)

and $\beta_1(x) = \mu_3^2(x)/\mu_2^3(x)$, $\beta_2(x) = \mu_4(x)/\mu_2^2(x)$.

Hence for $\rho > 0$, as x increases, γ' decreases, so that both $E(Y|x)$ and $\mathrm{var}(Y|x)$ increase with x. On the other hand, since $\beta_1(x)$ and $\beta_2(x)$ are constant, the conditional distributions are all of the same shape.

Case II. Let the marginal p.d.f. of Y be an S_B curve. In this case, the moments cannot be simplified, but from Johnson's table, E.S. Pearson (1962) has noted the behaviour of the conditional moment curves. $E(Y|x)$ is an increasing function of x, while at the transition point $x_0 = \gamma_2/\rho$ the distribution is symmetric, $\mathrm{var}(Y|x)$ has a maximum and $\beta_1(x)$ and $\beta_2(x)$ have minima but are otherwise monotonic. Variation of this kind in the scedastic and beta curves is present in most of the examples used by Pretorius (1930).

If $|\delta'| \geqslant 1/\sqrt{2}$ then we know that the S_B distributions are unimodal, so that if $\rho^2 \geqslant 1 - 2\delta_2^2$ then the conditional distributions are certainly unimodal.

Case III. Let the marginal p.d.f. be an S_U curve. We have from the univariate results:

$$E(Y|x) = -\omega \sinh \Omega, \quad \mathrm{var}(Y|x) = \tfrac{1}{2}(\omega^2 - 1)(\omega^2 \cosh 2\Omega + 1), \tag{4.2.9}$$

$$\mu_3(x) = -\tfrac{1}{4}\omega^2(\omega^2 - 1)^2\{\omega^2(\omega^2 + 2)\sinh 3\Omega + 3\sinh\Omega\},$$

$$\mu_4(x) = \tfrac{1}{8}(\omega^2 - 1)^2\{\omega^4(\omega^8 + 2\omega^6 + 3\omega^4 - 3)\cosh 4\Omega + 4\omega^4(\omega^2 + 2)\cosh 2\Omega + 3(2\omega^2 + 1)\},$$

where $\omega = \exp(1/2\delta'^2)$, $\Omega = \gamma'/\delta'$. (4.2.10)

The beta curves can be obtained. For this distribution, $\gamma' < 0$ implies the distribution is positively skew. If $\gamma' < 0$ and $\rho > 0$, Ω will decrease with x. Hence, from (4.2.9), $E(Y|x)$ is an increasing function of x, while $\mathrm{var}(Y|x)$ has a maximum at $x_0 = \gamma_2/\rho$ but is otherwise monotonic. From Johnson's table (1949a), the behaviour of the beta curves is similar to $\mathrm{var}(Y|x)$.

4.2d Median regression

The median value $\tilde{y}_x$ must satisfy the equation $K(Y|x) = \frac{1}{2}$. Since Y_0 has a normal distribution with mean ρx_0, we immediately have, from (4.2.1),

$$\gamma_2 + \delta_2 T_J(\tilde{y}_x) = \rho\{\gamma_1 + \delta_1 T_I(x)\},$$

so that the median regression curve of Y on X is given by

$$T_J(y) = \{(\rho\gamma_1 - \gamma_2)/\delta_2\} + (\rho\delta_1/\delta_2)T_I(x). \quad (4.2.11)$$

The regression is linear if $\rho\delta_1 = \delta_2$, $\rho\gamma_1 = \gamma_2$ and $T_J \equiv T_I$, which would imply serious restrictions on the parameters. The only other case is $T_J(x) \equiv T_I(x) = x$ which implies that the distribution is bivariate normal.

For the S_{BB} system we could not obtain $E(Y|x)$ explicitly, but from (4.2.11) the median regression is

$$y = \frac{\theta x^{\phi}}{(1-x)^{\phi} + \theta x^{\phi}},$$

where $\theta = \exp\{(\rho\gamma_1 - \gamma_2)/\delta_2\}$, $\phi = \rho\delta_1/\delta_2$.

Johnson (1949b) has given diagrams showing the general appearance of the curves given by (4.2.11). The general slope of the curve is found to depend upon the range of values of ρ.

4.2e A method of fitting

The systems depend on nine parameters; four in each marginal distribution and the parameter ρ which is $\mathrm{corr}(X_0, Y_0)$. A simple method of fitting an S_{IJ} surface consists in first fitting each marginal by a curve from an S_I system and then estimating ρ by the observed correlation coefficient between the transformed variables.

Example 4.1. Consider the fitting of an S_{IJ} system to the bean data of Example 3.1.

Johnson (1949a) has fitted curves of the system S_U to the distributions of the length (X) and the breadth (Y), respectively. It is found that

$$\gamma_1 = 2{\cdot}38, \quad \delta_1 = 2{\cdot}64, \quad \mu = 16{\cdot}0745, \quad \lambda_1 = 1{\cdot}5192,$$

$$\gamma_2 = 2{\cdot}13, \quad \delta_2 = 3{\cdot}55, \quad \nu = 8{\cdot}6195, \quad \lambda_2 = 0{\cdot}9721,$$

where μ, λ_1 are respectively the location and scale parameters of X and ν, λ_2 are those of Y. The correlation coefficient of the transformed variables

$$\sinh^{-1}\left(\frac{\text{length} - 16{\cdot}0745}{1{\cdot}5192}\right) \quad \text{and} \quad \sinh^{-1}\left(\frac{\text{breadth} - 8{\cdot}6195}{0{\cdot}9721}\right)$$

is found to be 0·746, and so we put $\rho = 0{\cdot}746$. From these values of the parameters, $E(Y|x)$ and $\mathrm{var}(Y|x)$ can be calculated using (4.2.9). The results are shown in Fig. 4.1–4.4. From Fig. 4.1 and 4.3, we observe that the regression curves are quite close fits, but Fig. 4.2 and

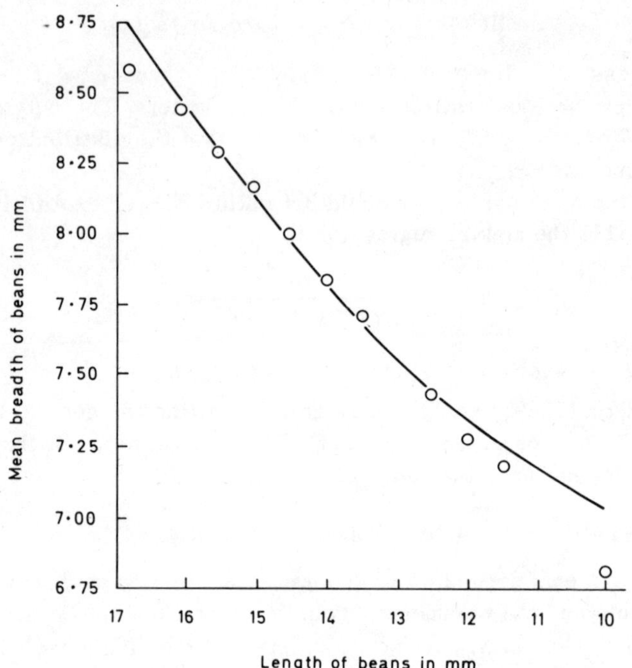

Fig.4.1 **Regression curve of breadth on length of beans (S_{UU} distribution)**

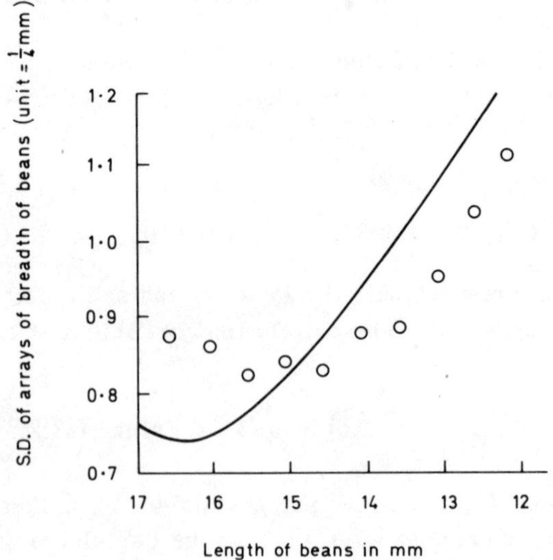

Fig.4.2 **Scedastic curve of breadth on length of beans (S_{UU} distribution)**

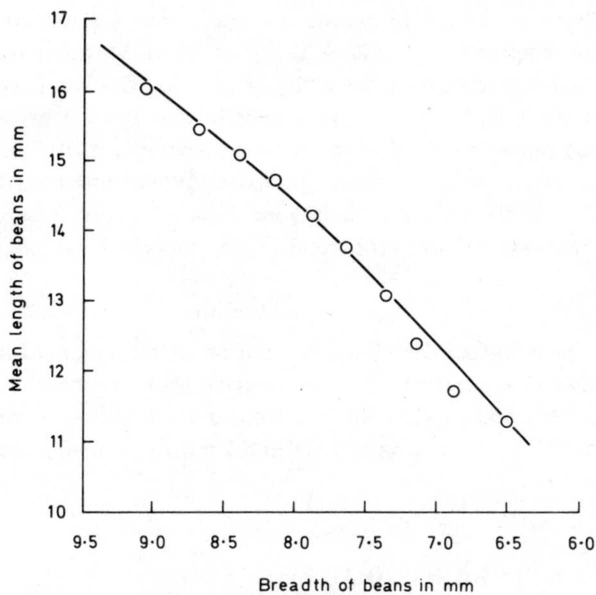

Fig.4.3 Regression curve of length on breadth of beans (S_{UU} distribution)

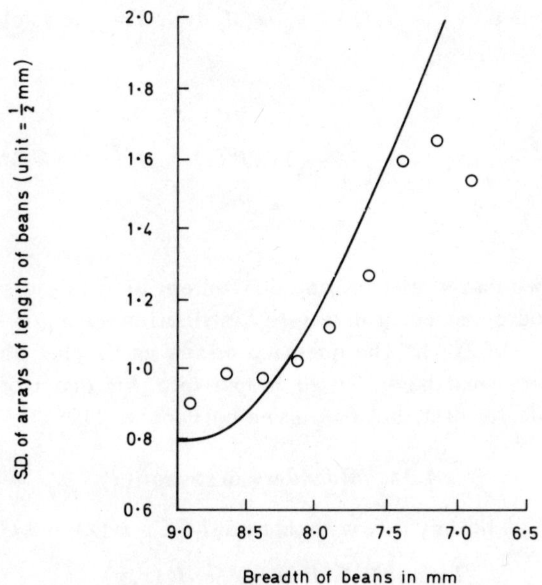

Fig.4.4 Scedastic curve of length on breadth of beans (S_{UU} distribution)

4.4 show that the scedastic curves are not giving a close fit. On comparing these diagrams with Fig. 3.1–3.4 of Type AA curves, we find that these scedastic curves are at least as good fits as Type AA curves and are certainly free from arbitrary oscillation. It must be noted that the Type AA surface contains fourteen parameters, while the S_U surface depends on only nine parameters. Johnson (1949b) has also fitted the beta curves, but the fit gets much more unsatisfactory, although the curves do indicate the general trend of the observed values.

4.3 An extension

Since the marginal distributions can be fitted successfully by various univariate systems, we may assume that the form of the marginal d.f's $F(x)$ and $G(y)$ is known. We can then define a continuous bivariate distribution having the marginal p.d.f's f and g by using the transformation

$$F(x) = B(x_0), \quad G(y) = B(y_0), \tag{4.3.1}$$

where $B(.)$ is the d.f. of $N(0, 1)$. Let

$$J_1 = B^{-1}F, \quad J_2 = B^{-1}G, \tag{4.3.2}$$

so that J_1 and J_2 are the transformations specified by the marginal distributions, i.e. $x_0 = J_1(x)$, $y_0 = J_2(y)$. The p.d.f. of (X, Y) is

$$h(x, y; \rho)$$

$$= \frac{f(x)\, g(y)}{\sqrt{(1-\rho^2)}} \exp\left[-\frac{\rho}{2(1-\rho^2)}\{\rho(J_1^2(x) + J_2^2(y)) - 2J_1(x)J_2(y)\}\right]. \tag{4.3.3}$$

In fact, we can start from any convenient bivariate distribution and by this method construct a bivariate distribution with given margins F and G (Nataf, 1962). But the question arises as to what characteristics the end-product must have. We go deeper into this question after studying the bounds for distributions given by Fréchet (1951).

4.3a Boundary distributions

Let $H(x,y)$ be any d.f. with marginal d.f's $F(x)$ and $G(y)$; we show that

$$H_{-1}(x,y) \leqslant H(x,y) \leqslant H_1(x,y), \tag{4.3.4}$$

where

$$H_1(x,y) = \min\{F(x), G(y)\} \tag{4.3.5}$$

and $$H_{-1}(x,y) = \max\{F(x) + G(y) - 1, 0\}. \qquad (4.3.6)$$

Further, $H_{-1}(x,y)$ and $H_1(x,y)$ are two d.f's with marginal d.f's F and G.

We have

$$H(x,y) \leqslant H(x,\infty) = F(x),\ H(x,y) \leqslant H(\infty,y) = G(y),$$

so that

$$H(x,y) \leqslant \min\{F(x), G(y)\} = H_1(x,y).$$

Since

$$\mathrm{P}(X > x, Y > y) = 1 - F(x) - G(y) + H(x,y),$$

we have

$$H(x,y) \geqslant F(x) + G(y) - 1$$

in addition to $H(x,y) \geqslant 0$, and so

$$H(x,y) \geqslant \max\{F(x) + G(y) - 1, 0\} = H_{-1}(x,y).$$

Hence (4.3.4) is established. We now proceed to show that $H_{-1}(x,y)$ and $H_1(x,y)$ are d.f's with margins F and G. Let us first consider $H_1(x,y)$. We verify that it satisfies the following sufficient conditions to be a d.f.:

(i) $H(\infty, \infty) = 1$;

(ii) $H(-\infty, y) = H(x, -\infty) = 0$;

(iii) $H(x + 0, y) = H(x, y + 0) = H(x,y)$; and

(iv) $H(a,b) + H(a + h, b + k) - H(a + h, b) - H(a, b + k) \geqslant 0$

holds for all $h, k \geqslant 0$.

We can write (4.3.5) as

$$H_1(x,y) = \tfrac{1}{2}\{F(x) + G(y) - |F(x) - G(y)|\}. \qquad (4.3.7)$$

Since F and G are d.f's, the first three conditions (i)–(iii) are easily seen to be satisfied. To prove that condition (iv) holds, we assume for simplicity that F and G are continuous, so that there exists a transformation $x' = F(x)$, $y' = G(y)$. Consequently, we can take (4.3.7) as

$$H_{-1}(x,y) = \tfrac{1}{2}\{x + y - |x - y|\}, \quad 0 \leqslant x,\ y \leqslant 1,$$

which implies that the condition (iv) is equivalent to

$$|c + h| + |c - k| \geqslant |c + h - k| + |c| \quad \text{for } h,k \geqslant 0,$$

where $c = a - b$. On squaring both sides, we see that this inequality holds if $$|c(c+h-k)-hk| \geqslant |c(c+h-k)| - |hk|$$

for $h,k \geqslant 0$, which is obviously true. Hence $H_1(x,y)$ is a d.f.

Similarly, we can write (4.3.6) as

$$H_{-1}(x,y) = \tfrac{1}{2}\{F(x)+G(y)-1+|F(x)+G(y)-1|\}. \quad (4.3.8)$$

Again, for simplicity, we transform the r.v's by $X' = F(X)$ and $Y' = G(Y)$, so that

$$H_{-1}(x,y) = \tfrac{1}{2}\{x+y-1+|x+y-1|\}, \quad 0 \leqslant x,\ y \leqslant 1.$$

The conditions (i)–(iii) are easily seen to be satisfied. For $H_{-1}(x,y)$ to be a d.f., we need to prove that condition (iv) holds i.e. to show that

$$|d+h+k| + |d| - |d+h| - |d+k| \geqslant 0 \quad \text{for } h,k \geqslant 0,$$

where $d = a+b-1$. On squaring both sides, we find that the inequality holds if

$$|d(d+h+k)+hk| \leqslant |d|\,|d+h+k| + |hk| \quad \text{for } h,\ k \geqslant 0,$$

which is obviously true.

It can easily be seen that $H_1(x,y)$ degenerates on a non-decreasing curve

$$F(x) = G(y),$$

while $H_{-1}(x,y)$ degenerates on a non-increasing curve

$$F(x) + G(y) = 1.$$

For independent r.v's X and Y, we can write

$$H_0(x,y) = F(x)\,G(y).$$

We have from (4.3.4)

$$H_{-1}(x,y) \leqslant H_0(x,y) \leqslant H_1(x,y).$$

The above results are due to Fréchet (1951), but his proofs of (4.3.5) and (4.3.6) are different. Konijn (1959) has used mixtures of the above distributions to investigate the power of certain tests of independence. Fréchet (1958) gives the d.f.

$$H(x,y) = (1-a-b)H_0(x,y) + aH_1(x,y) + bH_{-1}(x,y) \quad (4.3.9)$$

for $a,b \geqslant 0$ and $a+b \leqslant 1$. The class of these d.f's with two parameters contains $H_{-1}(x,y)$, $H_0(x,y)$ and $H_1(x,y)$.

For problems in statistical procedures concerning tests of

independence, it is of interest to construct a family of bivariate distributions having specified margins and just one parameter to measure the degree of dependence. Furthermore, if for some values of the parameter this family contains H_{-1}, H_0 and H_1, then the parameter would be a reasonable measure of dependence, indicating perfect positive dependence when it gives H_1, perfect negative dependence when it gives H_{-1}, and independence when it gives H_0. In the next section we show that $h(x,y;\rho)$ defines such a family, and later on, in Chapter 8, we meet another family with these characteristics. In fact, a very simple one-parameter family with these characteristics is given by

$$\begin{aligned} H(x,y) &= \tfrac{1}{2}\theta^2(1-\theta)H_{-1}(x,y) + (1-\theta^2)H_0(x,y) \\ &\qquad + \tfrac{1}{2}\theta^2(1+\theta)H_1(x,y), \qquad |\theta| \leqslant 1, \end{aligned}$$

which is a particular case of (4.3.9). However, no meaningful derivation can be obtained for this family.

4.3b Properties

For $\rho = 0$, we see from $h(x,y;\rho)$ that X and Y are independent random variables. Let $H(x,y;\rho)$ be the d.f. of the r.v. (X,Y), so that

$$H(x,y;\rho) = B[J_1(x),\ J_2(y);\rho], \tag{4.3.10}$$

where $B(x,y;\rho)$ is the d.f. of $N(0,0,1,1,\rho)$. We now show that

$$H(x,y;-1) = H_{-1}(x,y) \tag{4.3.11}$$

and

$$H(x,y;1) = H_1(x,y); \tag{4.3.12}$$

that is, the class also contains Fréchet's boundary distributions. On using the results

$$B(x_0,y_0;-1) = \max\{B(x_0) + B(y_0) - 1, 0\}$$

and

$$B(x_0,y_0;1) = \min\{B(x_0), B(y_0)\}$$

in (4.3.10) together with (4.3.2), we obtain (4.3.11) and (4.3.12).

In passing, we note the important result by Lancaster (1957) that $|\mathrm{corr}(X,Y)| \leqslant |\mathrm{corr}(X_0, Y_0)| = \rho$, and that equality holds only if (X,Y) is bivariate normal. For a proof, the reader is referred to Kendall and Stuart (1967, pp. 568–9). Van Uven (1925a,b; 1926, 1929) has developed a graphical method to obtain the transformations $X_0 = t_1(X)$ and $Y_0 = t_2(Y)$ which will make $\mathrm{corr}(t_1(X), t_2(Y))$ a maximum, but, in fact, $t_1 \equiv J_1$ and $t_2 \equiv J_2$ can only satisfy these requirements theoretically.

4.3c Regression

The p.d.f. of $Y_0 = J_2(Y)$ given $x_0 = J_1(x)$ is $N[\rho x_0, 1-\rho^2]$. $E(Y|x)$ is complicated and cannot be handled analytically. However, the median regression of Y on X is

$$J_1(y) = \rho J_2(x). \tag{4.3.13}$$

The regression is linear if and only if $J_1 \equiv J_2$, which implies $F = G = B$, i.e. the bivariate normal case. We can study the form of (4.3.13) by using an approximation to B^{-1}. Let

$$\Psi(x) = 1/(1+e^{-x})$$

be the d.f. of the logistic distribution. It can be shown that $\max_x |\Psi(cx) - B(x)|$ is minimized at $c = 0{\cdot}5875$, and at this value of c, $\max_x |\Psi(cx) - B(x)| \leqslant 0{\cdot}01$, so that we can take

$$B(x) \doteq 1/(1+e^{-cx}), \tag{4.3.14}$$

where $c = 0{\cdot}5875$. This approximation was given by Haley (1952). Berkson (1951) proposed approximating $B(x)$ by $\Psi(x)$, i.e. taking $c = 1$.

Using (4.3.14) in (4.3.13), the median regression of Y on X is approximately given by

$$G(y) = \frac{\{F(x)\}^{\rho}}{\{F(x)\}^{\rho} + \{1-F(x)\}^{\rho}} \tag{4.3.15}$$

which does not depend on c. The general slope clearly depends on values of ρ and the curve passes through the median point $F = G = \frac{1}{2}$.

Since the form of F and G is known, it is also useful to consider the mean regression of F on G. Now

$$E[G(Y)|F(x)] = E[B(Y_0)|x_0] = B[\rho x_0/\sqrt{(2-\rho^2)}],$$

so that the mean regression of $G(Y)$ on $F(X)$ is

$$G(y) = B[\rho x_0/\sqrt{(2-\rho^2)}]. \tag{4.3.16}$$

Again, on using the approximation given by (4.3.14), (4.3.16) reduces to

$$G(y) = \frac{\{F(x)\}^{a}}{\{F(x)\}^{a} + \{1-F(x)\}^{a}}, \tag{4.3.17}$$

where $a = \rho/\sqrt{(2-\rho^2)}$. This curve is similar to the median regression of Y on X given by (4.3.15).

Moreover, (4.3.14) provides an approximation to the marginal d.f's in Johnson's system. For simplicity, we take $x_0 = T_I(x)$ for the S_I system. The d.f. $F_I(x)$ obeys the relation $F_I(x) = B\{T_I(x)\}$, so that from (4.3.14), approximately

$$F_I(x) = 1/[1 + \exp\{-cT_I(x)\}]. \qquad (4.3.18)$$

Consequently, $F_L(x) = 1/(1 + x^{-c})$, $F_B(x) = 1/[1 + \{(1 - x)/x\}^c]$ and $F_U(x) = 1/[1 + \{x + \sqrt{(x^2 + 1)}\}^{-c}]$. Further, the d.f's can easily be inverted. For example,

$$F_U^{-1}(x) = \tfrac{1}{2}\,\frac{x^{2c} - (1-x)^{2c}}{\{x(1-x)\}^c}.$$

4.3d Random sampling

A random pair (x,y) from the population with p.d.f. (4.3.3) can easily be constructed from a random uniform pair (u,v) as follows. Let X_0^* and Y_0^* be independent $N(0,1)$. By Box and Muller (1958), a random pair (x_0^*, y_0^*) from this population is given by

$$x_0^* = (-\log u)^{1/2}\cos(2\pi v), \quad y_0^* = (-\log u)^{1/2}\sin(2\pi v). \qquad (4.3.19)$$

Now, if (X_0, Y_0) is $N(0,0,1,1,\rho)$ then

$$x_0 = x_0^*, \quad y_0 = \rho x_0^* + (1 - \rho^2)^{1/2} y_0^*,$$

so that from (4.3.1), a random pair (x,y) is

$$x = F^{-1}[B(x_0^*)], \quad y = G^{-1}[B\{\rho x_0^* + (1 - \rho^2)^{1/2} y_0^*\}], \qquad (4.3.20)$$

where x_0^* and y_0^* are given by (4.3.19). We can also use the approximation for $B(.)$ given by (4.3.14).

4.3e Applications

The fitting of this distribution can be done as in Johnson's systems, but with the obvious modification that the marginal distributions may be fitted by any univariate system. However, it is too much to expect that a system can provide an adequate description of the internal relationship between X and Y when only the parameter ρ is available for given margins. For robustness studies, our interest generally lies in certain broad types of departure from normality and the system can certainly serve the purpose. At least, it is free from the defects of the Edgeworth expansion.

Consider the distributions of Spearman's rank correlation coefficient, Kendall's τ, and the Fisher–Yates coefficient in sampling from bivariate normal populations. Since these coefficients are invariant under monotonic transformations applied to the marginal distributions, the distributions are the same for samples drawn from populations belonging to the above class. This fact has been noted by Fieller, Hartley and Pearson (1957). They have examined by Monte Carlo methods the adequacy of Fisher's z-transformation applied to the three coefficients in sampling from bivariate normal populations.

CHAPTER 5

A DEPENDENCE MODEL

5.1 Introduction

In Chapter 4 we generated distributions from a given bivariate distribution by the translation method. We can also construct bivariate distributions by transforming independent random variables. The resulting distributions will be models of departure from independence of the given variables. This approach is due to Steffensen (1922).

Let X_0 and Y_0 be independent random variables with given marginal p.d.f's $f_0(x)$ and $g_0(y)$ respectively. A simple dependence model can be constructed by the linear transformations

$$X = a_1 X_0 + b_1 Y_0 + c_1, \quad Y = a_2 X_0 + b_2 Y_0 + c_2. \tag{5.1.1}$$

Since f_0 and g_0 are known, there is no loss of generality in taking the inverse transformation of (5.1.1) as

$$X_0 = X + aY, \quad Y_0 = Y + bX. \tag{5.1.2}$$

The p.d.f. of (X, Y) is given by

$$h(x, y) = |ab - 1| \, f_0(x + ay) \, g_0(y + bx). \tag{5.1.3}$$

5.2 Properties

If $a = b = 0$, then obviously the random variables X and Y are independent. In fact, if the random variables are independent and the transformation is non-trivial (a, b and $|ab - 1| \neq 0$) then by a characterization of the normal distribution given in Feller (1966, pp. 77–80) we find that X, Y, X_0 and Y_0 are independent normally distributed random variables. So that if the distributions of X_0 and Y_0 are both non-normal then the random variables X and Y are dependent except for the trivial cases.

We now study the behaviour of the correlation coefficient of X and Y:

$$\rho = -\frac{a\bar{\sigma}_2^2 + b\bar{\sigma}_1^2}{\{(\bar{\sigma}_2^2 + b^2\bar{\sigma}_1^2)(\bar{\sigma}_1^2 + a^2\bar{\sigma}_2^2)\}^{1/2}}, \tag{5.2.1}$$

where $\bar{\sigma}_1^2 = \mathrm{var}(X_0)$ and $\bar{\sigma}_2^2 = \mathrm{var}(Y_0)$. If a and b are of the same sign we see that $\rho = 0$ only if $a = b = 0$. Hence in this case $\rho = 0$ implies that the r.v's X and Y are independent. Furthermore, $\rho^2 = 1$

if $ab = 1$, and from (5.1.2) it follows that the distribution degenerates on a line. These results have been noted by Steffensen (1941).

Another transformation of interest is the orthogonal transformation (Konijn, 1957):

$$X_0 = \cos\theta\, X - \sin\theta\, Y, \quad Y_0 = \sin\theta\, X + \cos\theta\, Y. \tag{5.2.2}$$

From (5.2.1), we easily see that in this case ρ depends on $(\bar{\sigma}_2^2 - \bar{\sigma}_1^2)$ $\tan\theta$. Hence $\rho = 0$ if $\bar{\sigma}_1 = \bar{\sigma}_2$ irrespective of independence, and if $\sigma_1 \neq \sigma_2$, $\rho = 0$ implies $\theta = 0$, so that X and Y are necessarily independent. For higher moments it is useful to note that

$$\mathrm{E}\{(X + aY)^r (Y + bX)^s\} = \mathrm{E}(X + aY)^r\ \mathrm{E}(Y + bX)^s. \tag{5.2.3}$$

Finally, we note that this distribution includes the bivariate normal (§ 10.6) and Rhodes' distribution (§ 10.16). In the latter case, X_0 and Y_0 have gamma distributions.

5.3 A method of fitting

Let us assume that X and Y are measured from their means. From (5.2.3),

$$\mathrm{E}\{(X + aY)^r (Y + bX)^s\} = 0 \quad \text{if } r \text{ or } s = 1,$$

and on putting $(r, s) = (1, 1), (1, 2)$ and $(2, 1)$, we have

$$b\mu_{20} + (1 + ab)\mu_{11} + a\mu_{02} = 0, \tag{5.3.1}$$

$$b^2\mu_{30} + b(2 + ab)\mu_{21} + (1 + 2ab)\mu_{12} + a\mu_{03} = 0, \tag{5.3.2}$$

$$b\mu_{30} + (1 + 2ab)\mu_{21} + a(2 + ab)\mu_{12} + a^2\mu_{03} = 0. \tag{5.3.3}$$

We need to solve these equations for a and b. Let us write

$$c = b/(1 + ab), \quad d = a/(1 + ab) \tag{5.3.4}$$

so that

$$a = \{1 - \surd(1 - 4cd)\}/2c, \quad b = \{1 - \surd(1 - 4cd)\}/2d.$$

On substituting these values in (5.3.1) – (5.3.3), we have

$$c\mu_{20} + d\mu_{02} + \mu_{11} = 0, \tag{5.3.5}$$

$$c\mu_{21} + d\mu_{03} + \mu_{12} = 0, \quad c\mu_{30} + d\mu_{12} + \mu_{21} = 0. \tag{5.3.6}$$

We can combine equations (5.3.6) by making

$$(c\mu_{21} + d\mu_{03} + \mu_{12})^2 + (c\mu_{30} + d\mu_{12} + \mu_{21})^2$$

a minimum. This gives

$$c\{\mu_{02}(\mu_{30}^2 + \mu_{21}^2) - \mu_{20}(\mu_{03}\mu_{21} + \mu_{30}\mu_{12})\}$$
$$+ d\{\mu_{02}(\mu_{03}\mu_{21} + \mu_{30}\mu_{12}) - \mu_{20}(\mu_{03}^2 + \mu_{12}^2)\}$$
$$+ \{\mu_{02}\mu_{21}(\mu_{30} + \mu_{12}) - \mu_{20}\mu_{12}(\mu_{03} + \mu_{21})\} = 0. \quad (5.3.7)$$

On using (5.3.5) and (5.3.7), a and b can be determined. Further, forming the variables $X_0 = X + aY$ and $Y_0 = Y + bX$, f_0 and g_0 can be fitted by any univariate system. In fact, the central moments $\bar{\mu}_r$ and $\bar{\nu}_s$ of X_0 and Y_0 can be obtained from

$$\bar{\mu}_r = \mu_{r0} + \binom{r}{1} a\,\mu_{r-1,1} + \binom{r}{2} a^2 \mu_{r-2,2} + \ldots + a^r \mu_{0r}, \quad (5.3.8)$$

$$\bar{\nu}_s = \mu_{0s} + \binom{s}{1} b\,\mu_{1,s-1} + \binom{s}{2} b^2 \mu_{2,s-2} + \ldots + b^s \mu_{s0}. \quad (5.3.9)$$

We now illustrate this method of fitting.

Example 5.1. In Table 5.1 the upper entries show a distribution of Müllerian glands on the right and left forelegs of 2000 male pigs. The number of glands on the right legs is denoted by X, and on the left legs by Y. It is found that the moments are

$$\begin{array}{lll} \mu = 3{\cdot}54650 & \nu = 3{\cdot}53950 & \mu_{11} = 2{\cdot}35366 \\ \mu_{20} = 2{\cdot}95684 & \mu_{02} = 2{\cdot}99444 & \mu_{21} = 1{\cdot}38461 \\ \mu_{30} = 2{\cdot}15954 & \mu_{03} = 2{\cdot}38197 & \mu_{12} = 1{\cdot}47138 \\ \mu_{40} = 25{\cdot}29494 & \mu_{04} = 27{\cdot}46311 & \mu_{22} = 18{\cdot}60165 \end{array}$$

$$\mu_{31} = 19{\cdot}37609, \; \mu_{13} = 20{\cdot}26674.$$

On using these moments in (5.3.5) and (5.3.7), we find that

$$c = -0{\cdot}414360, \; d = -0{\cdot}376853,$$

so that from (5.3.4),

$$a = -0{\cdot}467360, \; b = -0{\cdot}513874.$$

Consequently, from (5.3.8) and (5.3.9), we have

$$\begin{array}{ll} \bar{\mu}_2 = 1{\cdot}410889 & \bar{\nu}_2 = 1{\cdot}356272 \\ \bar{\mu}_3 = 0{\cdot}939207 & \bar{\nu}_3 = 0{\cdot}917503 \\ \bar{\mu}_4 = 6{\cdot}48562 & \bar{\nu}_4 = 6{\cdot}52408 \end{array}$$

So for the distribution of X_0, we have $\beta_1 = 0{\cdot}31$, $\beta_2 = 3{\cdot}26$ and for the distribution of Y_0, $\beta_1 = 0{\cdot}34$, $\beta_2 = 3{\cdot}55$. From these values it is seen that we may assume Pearson's Type III curves for both distributions. We therefore take

$$f_0(x) = c_1 (x_0 + A_1)^{p_1} e^{-m_1 x_0},$$
$$g_0(y) = c_2 (y_0 + A_2)^{p_2} e^{-m_2 y_0},$$

and using the moments $\bar{\mu}_r$ and $\bar{\nu}_r$ calculated above we find the following values for the constants:

$\log_{10} c_1 = \bar{8}\cdot16221,\quad A_1 = 4\cdot23891,\quad p_1 = 11\cdot73547,\quad m_1 = 3\cdot00442.$

$\log_{10} c_2 = \bar{8}\cdot98515,\quad A_2 = 4\cdot00974,\quad p_2 = 10\cdot85454,\quad m_2 = 2\cdot95644.$

Hence, from (5.1.3),

$$h(x, y) = |ab - 1| (x + ay + A_1)^{p_1} (y + bx + A_2)^{p_2} \times \exp\{-m_1 (x + ay) - m_2 (y + bx)\},$$

where X and Y have zero means. On transforming the variables so that X and Y have means μ and ν, we get

$$h(x,y) = (a_0 - a_1 y + x)^{p_1} (b_0 - b_1 x + y)^{p_2} \exp(-d_0 - d_1 x - d_2 y),$$

where the values of the constants are

$a_0 = 2\cdot34663$	$b_0 = 2\cdot29269$	$d_0 = 23\cdot71257$
$a_1 = 0\cdot467360$	$b_1 = 0\cdot513874$	$d_1 = 1\cdot48519$
$p_1 = 11\cdot73547$	$p_2 = 10\cdot85454$	$d_2 = 1\cdot55229$

This is in fact Rhodes' distribution (§ 10.16). Since the observed distribution is discrete, $h(x, y)$ may be regarded as the probability function, and the expected frequencies then obtained are shown in the lower entries of Table 5.1. Jørgensen (1916) treated the same example to fit the Edgeworth expansion, and in comparison the above fit is much better (Steffensen, 1922). Negative frequencies arise in several of the empty cells in fitting the Edgeworth expansion.

It may be noted that by examining the contour lines of some skew surfaces, Karl Pearson inferred much earlier that in most cases $z = h(x, y)$ could not be expressed in the form $z = f_0(x) \times g_0(y)$ by a rotation of axes, and a historical account is given in K. Pearson (1923a).

5.4 Other applications

In addition to the above application, the model has appeared in various statistical procedures. Haldane (1949) considered the effect of non-normality on the correlation coefficient for samples from

$$X = \frac{\sigma_1}{\sqrt{2}}[(1+\rho)^{\frac{1}{2}}X_0 + (1-\rho)^{\frac{1}{2}}Y_0], \quad Y = \frac{\sigma_2}{\sqrt{2}}[(1+\rho)^{\frac{1}{2}}X_0 - (1-\rho)^{\frac{1}{2}}Y_0].$$

Konijn (1956) has studied the Pitman efficiency of the tests of independence against the alternatives of the form

$$X = \lambda_1 X_0 + \lambda_2 Y_0, \quad Y = \lambda_3 X_0 + \lambda_4 Y_0,$$

the null hypothesis being $\lambda_1 = \lambda_4 = 1$, $\lambda_2 = \lambda_3 = 0$. Chatterjee (1960) has given sequential procedures for the regression parameters of the population with p.d.f.

$$\sigma^{-1} f_0(x)\, b\{(y - \beta x)/\sigma\},$$

where $b(.)$ is the p.d.f. of $N(0, 1)$ and $f_0(.)$ is unspecified. For the model

$$X = X_0, \quad Y = a_2 X_0 + b_2 Y_0,$$

Jogdeo (1964) considers power properties of certain non-parametric procedures for regression.

Table 5.1 Observed frequencies (*upper entry*)
(*lower entry*) of Müllerian glands on the
(X = the number of glands on the right legs

Y ↓	X→ 0	1	2	3	4
0	8 19	4 17	2 2		
1	5 15	151 78	65 63	14 13	5 1
2	2 2	58 58	154 158	88 114	27 30
3		9 13	96 108	173 187	119 121
4		3 1	28 30	128 117	153 149
5			7 4	28 36	77 86
6			1 –	6 6	26 28
7				– 1	3 6
8					1 1
9					
10					
Totals	15 36	225 167	353 365	437 474	411 422

and the corresponding expected frequencies
right and left forelegs of 2000 male pigs
and Y = the number of glands on the left legs)

5	6	7	8	9	10	Totals
						14
						38
1						241
–						170
7						336
3						365
24	8	1				430
37	5	–				471
92	16	8	1			429
88	29	5	1			420
101	58	20	3	1		295
87	48	16	3	–		280
52	48	18	5	3		159
47	39	20	7	2		149
11	16	17	3	3	–	53
16	20	15	7	2	1	68
9	7	9	2	2	–	30
4	7	7	5	2	1	27
–	–	5	2	2	1	10
1	2	2	2	1	1	9
	2	–	–	1		3
	–	1	1	1		3
297	155	78	16	12	1	2000
283	150	66	26	8	3	2000

CHAPTER 6

A REGRESSION MODEL

6.1 Definition

The families considered so far have at the most had specified marginal distributions, and the regression and scedastic curves have been incidental. Narumi (1923a,b) attacked the problem by specifying the regression and scedastic curves. He considered bivariate distributions with p.d.f's of the form

$$\left.\begin{aligned} h(x,y) &= \Phi_1(x)\,\Psi_1\,[\{y-u_1(x)\}/U_1(x)] \\ &= \Phi_2(y)\,\Psi_2\,[\{x-u_2(y)\}/U_2(y)]. \end{aligned}\right\} \tag{6.1.1}$$

The conditional p.d.f. of Y given X is

$$k(y|x) = \text{const.}\ \Psi_1[\{y-u_1(x)\}/U_1(x)],$$

so that the location parameter of Y given X is $u_1(x)$ and $U_1(x)$ is the scale parameter. Similarly, $u_2(y)$ and $U_2(y)$ are the location and scale parameters of the conditional distribution of X given Y. Hence we can describe $y = u_1(x)$ and $x = u_2(y)$ as the regression curves of Y on X and of X on Y, respectively. The corresponding scedastic curves are $y = U_1(x)$ and $x = U_2(y)$. Of course, in most cases, U_1^2 and U_2^2 will be equivalent to the conditional variances.

6.2 Solutions

We first outline a general method of solving these equations – an approach common to the particular cases dealt with by Narumi (1923a,b). Let us write

$$\psi_i = \log \Psi_i\,, \quad \phi_i = \log \Phi_i\,, \quad i = 1, 2, \tag{6.2.1}$$

so that (6.1.1) reduces to

$$\left.\begin{aligned} \log\ h(x,y) &= \phi_1(x) + \psi_1[\{y-u_1(x)\}/U_1(x)] \\ &= \phi_2(y) + \psi_2[\{x-u_2(y)\}/U_2(y)]. \end{aligned}\right\} \tag{6.2.2}$$

On differentiating (6.2.2) with respect to x and y, we get

$$\begin{aligned} &\{U_2(y)\}^2[\{u_1'(x) + uU_1'(x)\}\psi_1''(u) + U_1'(x)\psi_1'(u)] \\ &= \{U_1(x)\}^2\,[\{u_2'(y) + vU_2'(y)\}\,\psi_2''(v) + U_2'(y)\,\psi_2'(v)], \end{aligned} \tag{6.2.3}$$

where the dashes denote the derivative of the function with respect to its argument and

$$u = \frac{y - u_1(x)}{U_1(x)}, \quad v = \frac{x - u_2(y)}{U_2(y)}.$$

On solving the differential equation, we can obtain ψ_1 and ψ_2. These can be used to obtain ϕ_1 and ϕ_2 from the identity

$$\psi_1(u) - \psi_2(v) \equiv \phi_2(y) - \phi_1(x), \tag{6.2.4}$$

which follows from (6.2.2).

The following cases are of interest :-

Case I: Linear regression and constant scedasticity.
Case II: Linear regression and linear scedasticity.
Case III: Linear regression and parabolic variances.
Case IV: Regression and scedastic curves both equilateral hyperbolas.

We now proceed to investigate solutions for these cases.

Case I. Let us take

$$u_1 = m_1 x + c_1, \quad u_2 = m_2 y + c_2 \tag{6.2.5}$$

and U_1, U_2 constants so that (6.2.3) reduces to the form

$$m_1 \psi_1'' (y - m_1 x - c_1) = m_2 \psi_2'' (x - m_2 y - c_2). \tag{6.2.6}$$

On differentiating this relation with respect to x and with respect to y, we get

$$-m_1^2 \psi_1''' (y - m_1 x - c_1) = m_2 \psi_2''' (x - m_2 y - c_2)$$

and
$$m_1 \psi_1''' (y - m_1 x - c_1) = -m_2^2 \psi_2''' (x - m_2 y - c_2).$$

Except for the trivial cases, we have

$$\psi_1''' (y - m_1 x - c_1) = \psi_2''' (x - m_2 y - c_2) = 0.$$

Hence we have

$$\left.\begin{aligned} \psi_1(y - m_1 x - c_1) &= a_0 (y - m_1 x - c_1)^2 + a_1 (y - m_1 x - c_1) + a_2, \\ \psi_2(x - m_2 y - c_2) &= b_0 (x - m_2 y - c_2)^2 + b_1 (x - m_2 y - c_2) + b_2, \end{aligned}\right\} \tag{6.2.7}$$

so that both conditional distributions are bivariate normal.

Further, on substituting (6.2.7) in (6.2.4), we find

$$\phi_2(y) - \phi_1(x) \equiv a_0(y - m_1x - c_1)^2 + a_1(y - m_1x - c_1) + a_2 - b_0(x - m_2y - c_2)^2 - b_1(x - m_2y - c_2) - b_2.$$

But from (6.2.6) we have $a_0 m_1 = b_0 m_2$, so that the marginal distributions are also normal. Hence the distribution must be bivariate normal.

Case II. We assume u_1 and u_2 are given by (6.2.5) and take $U_1(x) = x + a_1$, $U_2(y) = y + a_2$; then (6.2.3) can be written as

$$\{k_1 + k_2(u + m_1)\}^2\{\psi_1'(u) + (u + m_1)\psi_1''(u)\} = \{k_2 + k_1(v + m_2)\}^2\{\psi_2'(v) + (v + m_2)\psi_2''(v)\}, \quad (6.2.8)$$

with the help of the inverse transformation of (u, v)

$$x + a_1 = \{k_2 + k_1(v + m_2)\}/S, \quad y + a_2 = \{k_1 + k_2(u + m_1)\}/S, \quad (6.2.9)$$

where $u = (y - m_1x - c_1)/(x + a_1)$, $\quad v = (x - m_2y - c_2)/(y + a_2)$ (6.2.10)

$$k_1 = m_1a_1 - c_1 - a_2, \quad k_2 = m_2a_2 - c_2 - a_1, \quad S = (u + m_1)(v + m_2) - 1.$$

Since the left-hand side of (6.2.8) is a function of u only and the right of v only, they must be equal to a constant. Hence, we have

$$\frac{d}{du}\{\psi_1'(u)(u + m_1)\} = \frac{c}{\{k_1 + k_2(u + m_1)\}^2}$$

so that $\quad (u + m_1)\psi_1'(u) = c_1 - c_2\{k_1 + k_2(u + m_1)\}^{-1}$,

where c_1 and c_2 are constants. Consequently,

$$\Psi_1(u) = c(u + m_1)^{p_2}\{k_1 + k_2(u + m_1)\}^q,$$

where c, p_2 and q are constants.

On substituting u from (6.2.10) and using (6.2.9), we have

$$\Psi_1\left(\frac{y - m_1x - c_1}{x + a_1}\right) = c\frac{(y + g_2)^{p_2}}{(x + a_1)^{p_2}}\{S(y + a_2)\}^q, \quad (6.2.11)$$

where $g_2 = m_1a_1 - c_1$. Similarly, with $g_1 = m_2a_2 - c_2$,

$$\Psi_2\left(\frac{x - m_2y - c_2}{y + a_2}\right) = c'\frac{(x + g_1)^{p_1}}{(y + a_2)^{p_1}}\{S(x + a_2)\}^q.$$

On using these values in (6.2.4), we have

$$\phi_1(x) - \phi_2(y) = \log c' + p_1 \log(x + g_1) + (p_2 + q) \log(x + a_1) - \log c - p_2 \log(y + g_2) - (p_1 + q) \log(y + a_2).$$

Accordingly,

$$\phi_1(x) = \log c'' + p_1 \log(x + g_1) + (p_2 + q) \log(x + a_1) = \log \Phi_1(x) \tag{6.2.12}$$

and $\phi_2(y)$ can be written similarly. Substituting (6.2.11) and (6.2.12) in (6.1.1), we thus have

$$h(x, y) = k_0(x + g_1)^{p_1}(y + g_2)^{p_2}(k_1 x + k_2 y)^q. \tag{6.2.13}$$

This form contains the bivariate beta (§ 10.9), the bivariate Pareto (§ 10.12) and the bivariate F (§ 10.14) distributions.

Further, if we take $U_1(x) = a_1$, $U_2(y) = y + a_2$, we find by the same process that

$$h(x, y) = k_0(ax + b)^{p_1} (c_1 x + c_2 y)^q e^{-ky}, \tag{6.2.14}$$

which contains McKay's bivariate gamma distribution (§ 10.10).

Case III. If $u_1(x) = m_1 x + c_1$, $U_1(x) = \sqrt{(x^2 + a_1)}$, etc., we find by the same method that

$$h(x, y) = k_0\{ax^2 + 2hxy + by^2 + 2gx + 2fy + c\}^n. \tag{6.2.15}$$

This contains the bivariate Cauchy (§ 10.8), the bivariate Student's t (§ 10.13) and the bivariate Type II (§ 10.15) distributions.

Case IV. If $u_1(x) = 1/(x + a)$, $U_1(x) = 1/(x + b)$, etc., we have by the same process

$$h(x, y) = k_0(x + f_1)^{\gamma_1} (y + g)^{\gamma_2} e^{\gamma(x+f_2)(y+g_1)},$$

which has both conditional distributions of the gamma type. This distribution has not appeared in practice.

Narumi (1923b) investigated fully the case of U_1 and U_2 both constants, but the other forms are of no practical interest (so far).

The underlying assumption of Narumi's class is that the β_1 and β_2 coefficients for the conditional distributions are constants. Pretorius (1930) found in his exhaustive analysis of six observed bivariate distributions that this assumption does not hold for these distributions.

Jogdeo (1968) has studied the family satisfying the conditions that the conditional d.f. $K(y \mid x)$ is monotonic in x and the d.f. $K(x \mid y)$ is monotonic in y. It can be seen that for U_1 and U_2 both constants, Narumi's class satisfies these conditions.

CHAPTER 7

CANONICAL REPRESENTATIONS

7.1 Introduction

We know that the p.d.f. of $N(0, 0, 1, 1, \rho)$ can be expressed as the tetrachoric series

$$b(x,y;\rho) = \left\{1 + \sum_{i=1}^{\infty} \rho^i H_i(x) H_i(y)\right\} b(x)\, b(y), \qquad (7.1.1)$$

where $b(.)$ is the p.d.f. of $N(0, 1)$ and $\{H_i(.)\}$ are the Hermite polynomials defined by (3.1.6), standardized so that

$$\int_{-\infty}^{\infty} H_i^2(x)\, b(x)\, dx = 1.$$

The sets $\{H_i(x)\}$ and $\{H_i(y)\}$ are complete sets of orthonormal functions on the marginal distributions $b(x)$ and $b(y)$. It is natural to enquire whether this type of expansion can be extended to any r.v. (X, Y) with given marginal d.f.'s F and G. The problem has been tackled in the general case by Lancaster (1958) from this angle.

7.2 A general expansion

Let (X, Y) be a random variable with specified marginal d.f.'s F and G. Let $\{x_i(X), y_i(Y)\}$ be complete sets of orthonormal functions on the marginal distributions $F(x)$ and $G(y)$, respectively. That is,

$$\int x_i(x)\, x_j(x)\, dF(x) = \int y_i(y)\, y_j(y)\, dG(y) = \begin{cases} 1, & \text{if } i = j \\ 0, & \text{if } i \neq j \end{cases}. \qquad (7.2.1)$$

Further, if $x' = x'(X)$ and $y' = y'(Y)$ are new variables with finite variances, then we may write

$$\left.\begin{aligned} X' &= a_0 + a_1 x_1(X) + a_2 x_2(X) + \dots \\ Y' &= b_0 + b_1 y_1(Y) + b_2 y_2(Y) + \dots \end{aligned}\right\} \qquad (7.2.2)$$

where Σa_i^2 and Σb_i^2 are convergent.

Let $H(x, y)$ be the d.f. of the r.v. (X, Y) and let

$$\rho_{ij} = \operatorname{corr}(x_i(X), y_j(Y)) = \iint x_i(x)\, y_j(y)\, dH(x, y). \qquad (7.2.3)$$

By the Cauchy–Schwarz inequality and (7.2.1), ρ_{ij} always exists and

$$\rho_{00} = 1, \quad \rho_{0k} = \rho_{k0} = 0. \qquad (7.2.4)$$

If $\sum_{i=1}^{\infty}\sum_{j=1}^{\infty}\rho_{ij}^2$ is convergent, then by some well-known results in the theory of integral equations (Courant and Hilbert, 1953, pp. 50–7), we have

$$dH(x, y) = \left\{ \sum_{i=0}^{\infty}\sum_{j=0}^{\infty} \rho_{ij}\, x_i\, y_j \right\} dF(x)\, dG(y). \tag{7.2.5}$$

For the proof the reader is referred to Lancaster (1958).

The quantity $\Sigma\Sigma\rho_{ij}^2$ can be related to the Karl Pearson coefficient of contingency ϕ^2 defined by

$$\phi^2 + 1 = \int (dH)^2/(dF\, dG) \equiv \int \Omega^2\, dF\, dG, \tag{7.2.6}$$

where Ω is the Radon–Nikodym derivative of $H(x, y)$ with respect to $F(x)\, G(y)$. On substituting (7.2.5) in (7.2.6) and using (7.2.1) and (7.2.3), we have

$$\phi^2 = \sum_{i=1}^{\infty}\sum_{j=1}^{\infty} \rho_{ij}^2\,. \tag{7.2.7}$$

Hence, if ϕ^2 is bounded for the r.v. (X, Y) then its p.d.f. can always be expressed in the form (7.2.5). It should be noted that for the bivariate normal distribution with correlation coefficient ρ, we have

$$\phi^2 = \rho^2/(1 - \rho^2).$$

The above representation (7.2.5) is due to Lancaster (1958). Rényi (1959) examines the adequacy of ϕ^2, given by (7.2.6), as a measure of dependence.

7.3 Canonical form

The expansion (7.2.5) is similar to the tetrachoric expansion (7.1.1) if $\rho_{ij} = 0$ for $i \neq j$. In general, it is possible to select the sets $\{x_i\}$ and $\{y_i\}$, so that $\rho_{ij} = 0$ if $i \neq j$. Let $\{\xi_i\}$ and $\{\eta_i\}$ be orthonormal sets on the marginal distributions. The pairs (ξ_1, η_1), (ξ_2, η_2), ... are chosen successively to maximize the correlations,

$$\rho_i = \text{corr}\,(\xi_i, \eta_i) = \int\int \xi_i\, \eta_i\, dH(x, y). \tag{7.3.1}$$

We shall show that these sets $\{\xi_i\}$ and $\{\eta_i\}$ satisfy the requirement corr $(\xi_i, \eta_j) = 0$ for $i \neq j$ and are subsets of $\{x_i\}$ and $\{y_i\}$ respectively. The variables (ξ_i, η_j) are called canonical variables and the ρ_i's are called canonical correlations.

Let $\xi_0 = \eta_0 = 1$ and let the orthogonal and normalizing conditions be

$$\int \xi_i \, dF(x) = \int \eta_i \, dG(y) = 0, \quad i = 1, 2, \dots \tag{7.3.2}$$

$$\int \xi_i \xi_j \, dF(x) = \int \eta_i \eta_j \, dG(y) = \begin{Bmatrix} 1, \text{ if } i = j \\ 0, \text{ if } i \neq j \end{Bmatrix}. \tag{7.3.3}$$

We first show that the canonical variables obey a second set of orthogonal conditions, namely

$$\iint \xi_i \eta_j \, dH(x, y) = 0 \quad \text{if } i \neq j. \tag{7.3.4}$$

Let us take $j > i$ for definiteness. By hypothesis, $\mathrm{E}(\xi_i \eta_i) = \rho_i$ is maximal in the recursive manner as described above. For establishing a contradiction to (7.3.4), it is sufficient to assume that $\mathrm{E}(\xi_i \eta_j)$ is not zero but equal to $\rho_i \tan\theta$. In view of (7.3.2) and (7.3.3), the function $\cos\theta\eta_i + \sin\theta\eta_j$ satisfies all necessary orthogonal and normalizing conditions. Now its correlation with ξ_i is $\rho_i \sec\theta$, but this is greater than ρ_i. Hence a contradiction results, so that (7.3.4) is true.

Now we show that the canonical variables $\{\xi_i\}$ and $\{\eta_i\}$ form subsets of the complete sets of orthonormal functions $\{x_i\}$ and $\{y_i\}$, so that (7.2.5) also holds for the canonical variables. In virtue of (7.3.3), $\mathrm{var}(\xi_i)$ and $\mathrm{var}(\eta_i)$ exist, and since the correlation is unaffected by a change of origin or scale, from (7.2.2) we may write

$$\xi_i = \sum_{k=1}^{\infty} a_{ik} x_k, \qquad \sum_{k=1}^{\infty} a_{ik}^2 = 1,$$

$$\eta_i = \sum_{k=1}^{\infty} b_{ik} y_k, \qquad \sum_{k=1}^{\infty} b_{ik}^2 = 1.$$

Now let us determine ξ_1 and η_1 so that

$$\mathrm{corr}(\xi_1, \eta_1) = \sum_{i=1}^{\infty}\sum_{j=1}^{\infty} a_{1i} b_{1j} \rho_{ij} \tag{7.3.5}$$

is maximum. Since $\sum_{i=1}^{\infty}\sum_{j=1}^{\infty} \rho_{ij}^2$ is convergent, from the theory of quadratic forms in infinitely many variables, we know that the bilinear form (7.3.5) has an attained maximum value at $a_{1i} = a_i$, $b_{1j} = b_j$, say. We can define a new set of variables

$$x_1^* = \xi_1 = \sum_{i=1}^{\infty} a_i x_i,$$

$$x_2^* = a_{21} x_1 + a_{22} x_2,$$

$$x_3^* = a_{31} x_1 + a_{32} x_2 + a_{33} x_3,$$

$$\cdots\cdots\cdots$$

where the a_{2i}, a_{3i}, ... are chosen to satisfy the orthogonal and normalizing conditions. A similar transformation can be applied to the y_i, so that

$$y_1^* = \eta_1 = \sum_{i=1}^{\infty} b_i y_i ,$$

$$y_2^* = b_{21} y_1 + b_{22} y_2 ,$$

$$y_3^* = b_{31} y_1 + b_{32} y_2 + b_{33} y_3 ,$$

$$\cdots\cdots$$

In virtue of (7.3.4), we now have

$$\rho_{1i}^* = \rho_{i1}^* = 0; \quad i \neq 1.$$

We can proceed similarly to find ξ_2 and η_2 in terms of the $\{x_i\}$ and $\{y_i\}$ respectively. Since ξ_2 is orthogonal to ξ_1, we can take

$$\xi_2 = \sum_{i=2}^{\infty} a_i^* x_i^* , \quad \sum_{i=2}^{\infty} a_i^{*2} = 1,$$

and

$$\eta_2 = \sum_{i=2}^{\infty} b_i^* y_i^* , \quad \sum_{i=2}^{\infty} b_i^{*2} = 1.$$

Further, $\{a_i^*\}$ and $\{b_i^*\}$ can be chosen so that $\sum_{i=2}^{\infty} \sum_{j=2}^{\infty} a_i^* b_j^* \rho_{ij}^*$ is a maximum. We again take a new set of variables

$$x_1^+ = x_1^* = \xi_1,$$

$$x_2^+ = \sum_{i=2}^{\infty} a_i^* x_i^* = \xi_2,$$

$$x_3^+ = a_{32}^* x_2^* + a_{33}^* x_3^*,$$

$$x_4^+ = a_{42}^* x_2^* + a_{43}^* x_3^* + a_{44}^* x_4^*,$$

$$\cdots\cdots$$

and similarly define y_1^+, y_2^+, y_3^+, ... in terms of the y_i. Now

$$\rho_{1i}^+ = \rho_{i1}^+ = 0; \quad i \neq 1; \quad \rho_{2i}^+ = \rho_{i2}^+ = 0; \quad i \neq 2.$$

We can proceed in an analogous way to obtain further canonical variables (ξ_3, η_3), (ξ_4, η_4), etc. from $\{x_i\}$ and $\{y_i\}$. Hence the result is proved, and consequently the probability element of (X, Y) from (7.2.5) is

$$dH(x, y) = \left\{ 1 + \sum_{i=1}^{\infty} \rho_i \, \xi_i \eta_i \right\} dF(x) \, dG(y), \tag{7.3.6}$$

provided $\phi^2 = \sum_{i=1}^{\infty} \rho_i^2$ is convergent. An alternative improved derivation of (7.3.6) is given by Lancaster (1963), but the above method of Lancaster (1958) brings out the statistical implications of (7.3.6) more clearly.

For discrete variables, (7.3.6) is the identity of Fisher as cited by Maung (1942) who amplified the canonical theory of Fisher (1940) for $m \times n$ contingency tables. For continuous variables, (7.2.5) and (7.3.6) were studied by Barrett and Lampard (1955), who also gave some interesting results when the r.v's X and Y are time-dependent. The above generalization and canonical interpretation of (7.3.6) is due to Lancaster (1958).

7.4 Properties

We now consider the converse of the result of §7.3. Let us assume that we are given the r.v. (X, Y) with p.d.f. (7.3.6). We then show that (ξ_i, η_i) is the ith pair of canonical variables. Suppose that (X', Y') is the first pair of canonical variables. We may write

$$X' = \sum_{i=1}^{\infty} a_i \xi_i, \quad \sum_{i=1}^{\infty} a_i^2 = 1, \tag{7.4.1}$$

$$Y' = \sum_{i=1}^{\infty} b_i \eta_i, \quad \sum_{i=1}^{\infty} b_i^2 = 1, \tag{7.4.2}$$

so that

$$\text{corr}\,(X', Y') = \sum_{i=1}^{\infty} a_i b_i \rho_i . \tag{7.4.3}$$

Writing $c_i = \rho_i a_i / \{ \sum_{i=1}^{\infty} (\rho_i a_i)^2 \}^{\frac{1}{2}}$, the equation (7.4.3) becomes

$$\text{corr}\,(X', Y') = \left\{ \sum_{i=1}^{\infty} (\rho_i a_i)^2 \right\}^{\frac{1}{2}} \left\{ \sum_{i=1}^{\infty} c_i b_i \right\}.$$

Now by the Cauchy–Schwarz inequality $\sum_{i=1}^{\infty} c_i b_i$ is less than 1 unless $b_i = c_i$. Hence the maximum of (7.4.3) with respect to the b_i's is

$$\left\{ \sum_{i=1}^{\infty} (\rho_i a_i)^2 \right\}^{\frac{1}{2}} = \left\{ \sum_{i=2}^{\infty} (\rho_i^2 - \rho_1^2)\, a_i^2 + \rho_1^2 \right\}^{\frac{1}{2}}$$

which is maximized by taking $a_i = 0$, $i = 2, 3, \ldots$. Consequently, (7.4.3) is maximized by taking $a_1 = b_1 = 1$ and all other coefficients zero. Hence from (7.4.1) and (7.4.2), the first pair of canonical variables is (ξ_1, η_1). Suppose now that (X'', Y'') is the second pair of canonical variables. Let these be standardized and put

$$X'' = \sum_{i=1}^{\infty} c_i \xi_i , \quad Y'' = \sum_{i=1}^{\infty} d_i \eta_i . \tag{7.4.4}$$

Since (X'', Y'') is uncorrelated with the first pair, we have $c_1 = d_1 = 0$ and

$$\text{corr}\,(X'', Y'') = \sum_{i=2}^{\infty} c_i\, d_i\, \rho_i \,.$$

The algebra used above shows that the corr (X'', Y'') is maximized when $c_2 = d_2 = 1$ and $c_i = d_i = 0$ $(i = 1, 3, 4, \ldots)$. Hence (ξ_2, η_2) is the second pair of canonical variables and the process can be continued.

We now give important members of (7.3.6). It can easily be seen from the tetrachoric expansion that the bivariate normal distribution (§ 10.6) belongs to this class. Further, in this case, the ith pair of canonical variables is $\{H_i(x), H_i(y)\}$ and this pair has canonical correlation $|\rho|^i$. We now show that the Wicksell–Kibble bivariate gamma distribution (§ 10.10) also belongs to this class. From the Hille–Hardy formula (Erdélyi *et al.*, 1953, p. 189) the p.d.f. of this distribution is

$$h_1(x, y) = \frac{C^{-\frac{1}{2}(p-1)}}{\Gamma(p)(1-C)} \exp\left\{-\frac{x+y}{1-C}\right\} (xy)^{\frac{1}{2}(p-1)}\, I_{p-1}\left(\frac{2\sqrt{(Cxy)}}{1-C}\right)$$

$$= \sum_{i=0}^{\infty} \frac{C^i\, i!\, \Gamma(p)}{\Gamma(i+p)} L_i^{(p)}(x)\, L_i^{(p)}(y), \quad x, y > 0, 0 < C < 1,$$

where $L_i^{(p)}(x)$ is the ith Laguerre polynomial given by

$$L_i^{(p)}(x) = \frac{e^x x^{-(p-1)}}{i!} \left(\frac{d}{dx}\right)^i (e^{-x} x^{i+p-1})$$

$$= \sum_{r=0}^{i} \binom{i+p-1}{i-r} \frac{(-x)^r}{r!},$$

and

$$\int_0^{\infty} e^{-x} x^{p-1} L_i^{(p)}(x)\, L_j^{(p)}(x)\, dx = \Gamma(i+p)/i! \quad \text{if } i = j,$$

$$= 0, \quad \text{if } i \neq j.$$

On standardizing $\{L_i^{(p)}(x)\}$ and $\{L_i^{(p)}(y)\}$ we find that

$$h_1(x, y) = \sum_{i=0}^{\infty} C^i L_i^{(p)}(x)\, L_i^{(p)}(y). \tag{7.4.5}$$

Hence the Wicksell–Kibble bivariate gamma distribution is of the form (7.3.6). Furthermore, the ith pair of canonical variables for this distribution is $\{L_i^{(p)}(x), L_i^{(p)}(y)\}$ and this pair has canonical correlation C^i. Barrett and Lampard (1955) gave these two expansions as examples of (7.3.6). They also studied the expansion (7.3.6) for a sine wave of random phase, and Leipnik (1959) has shown this expansion to be point-wise divergent.

Mehr (1965) gave a characterization of (7.3.6) when (i) $F(x) \equiv G(x)$, (ii) F is continuous, and (iii) the distribution of X is symmetrical about zero. Mehr (1968) has shown that Markov processes defined on (7.3.6) share some of the important properties of Gaussian processes.

We now obtain regression curves of the canonical variables. From (7.3.6), we have

$$\mathrm{E}(\eta_j|\xi_i) = \int \eta_j \, dG(y) + \sum_{i=1}^{\infty} \rho_i \xi_i \int \eta_i \eta_j \, dG(y),$$

so that from (7.3.2) and (7.3.3),

$$\begin{aligned} \mathrm{E}(\eta_j|\xi_i) &= \rho_i \xi_i, \quad \text{if } i = j, \\ &= 0, \quad \text{if } i \neq j. \end{aligned} \tag{7.4.6}$$

Hence the regressions of η_i on ξ_i and of ξ_i on η_i are linear, and for $i \neq j$ the regressions of η_j on ξ_i and of ξ_j on η_i are zero.

7.5 Applications

Lancaster (1958) has given a test of the goodness of fit for the bivariate normal distribution using this representation and has illustrated the method. This displays as components of χ^2, the contributions of the regressions of the ith Hermite polynomial in X on the jth polynomial in Y. Farlie (1961) studies the Pitman efficiency of a test of independence based on Daniels' generalized correlation coefficient against alternatives of the form (7.3.6). Lancaster and Hamdan (1964) use canonical representations to obtain an estimate of the correlation coefficient in the general $r \times c$ table, which is an extension of the tetrachoric method.

CHAPTER 8

CONTINGENCY-TYPE DISTRIBUTIONS

8.1 Definition

In § 4.3 we discussed the importance of constructing one-parameter families of bivariate distributions with given margins, which contain the Fréchet boundary distributions and the distribution corresponding to independent random variables. We study here another such class. We know that the coefficient of contingency ψ is defined by

$$\psi = p_1 p_4 / p_2 p_3$$

for the two-point distribution

$$p(x_1,y_1) = p_1, \quad p(x_1,y_2) = p_2, \quad p(x_2,y_1) = p_3, \quad p(x_2,y_2) = p_4.$$

Any bivariate distribution having d.f. H and marginal d.f's F and G can be dichotomized at an arbitrary point (x,y) and then, putting

$$p_1 = H, \quad p_2 = F - H, \quad p_3 = G - H, \quad p_4 = 1 - F - G + H,$$

we have

$$\frac{H(1 - F - G + H)}{(F - H)(G - H)} = \psi. \tag{8.1.1}$$

Using this idea, Plackett (1965) has constructed a class of bivariate distributions for given margins F and G, where the d.f. H is defined as that root of the equation

$$(\psi - 1)H^2 - \{1 + (F + G)(\psi - 1)H\} + \psi FG = 0, \quad \psi > 0, \tag{8.1.2}$$

which satisfies the Fréchet inequalities (discussed in § 4.3a)

$$H_{-1}(x,y) = \max(F + G - 1, 0) \leqslant H \leqslant \min(F,G) = H_1(x,y). \tag{8.1.3}$$

Following Mardia (1967a), we show that

$$H = [S - \{S^2 - 4\psi(\psi - 1)FG\}^{1/2}]/\{2(\psi - 1)\}, \quad \psi > 0, \tag{8.1.4}$$

where $\quad S = 1 + (F + G)(\psi - 1).$ $\qquad$ (8.1.5)

Suppose that α and β are the roots of the quadratic equation (8.1.2) and let β denote the root equal to the right-hand side of (8.1.4) and

$$\alpha = [S + \{S^2 - 4\psi(\psi - 1)FG\}^{1/2}]/\{2(\psi - 1)\}, \quad \psi > 0.$$

We first prove that α does not satisfy (8.1.3). This we do by treating the three cases $\psi > 1$, $0 \leqslant \psi < 1$ and $\psi = 1$ separately. If $\psi > 1$, we immediately find that

$$\alpha \geqslant \frac{1}{2(\psi - 1)} + \frac{1}{2}(F + G) \geqslant \frac{1}{2}(F + G) \geqslant \min(F, G).$$

Hence, in this case, α does not satisfy (8.1.3). Now, if $0 \leqslant \psi < 1$, the product of the roots is negative, so that the roots have opposite signs. But by considering the value of β with $0 \leqslant \psi < 1$, we find that β is the positive root, and so α, being negative, cannot satisfy (8.1.3). On taking these results together, we conclude that α does not satisfy (8.1.3) for $\psi \geqslant 0$, and hence must be discarded.

We now show that β (or H) is a d.f. by verifying that H satisfies the following conditions:

(i) $H(\infty, \infty) = 1$

(ii) $H(-\infty, y) = H(x, -\infty) = 0,$

(iii) $H(x + 0, y) = H(x, y + 0) = H(x, y)$ and

(iv) $H(a, b) + H(a + h, b + k) - H(a + h, b) - H(a, b + k) \geqslant 0$

holds for all $h, k \geqslant 0$.

Conditions (i)–(iii) are easy to verify. To verify condition (iv) we transform the r.v's so that F and G correspond to uniform variables on (0,1). Since the resulting H is continuous, condition (iv) is equivalent to showing that $\partial^2 H/\partial x\, \partial y$ exists and is positive, i.e. that

$$h_U(x, y; \psi) = \frac{\psi\{(\psi - 1)(x + y - 2xy) + 1\}}{[\{1 + (\psi - 1)(x + y)\}^2 - 4\psi(\psi - 1)xy]^{3/2}} \qquad (8.1.6)$$

exists and is positive where $0 \leqslant x, y \leqslant 1$. This is true, and consequently H given by (8.1.4) is a d.f.

It can be seen from (8.1.4) that for $G = 1$, $H = F$ and for $F = 1$, $H = G$. This result confirms that F and G are the marginal d.f's. Similarly, we find that as $\psi \to 1$, $H \to FG$. Further, as $\psi \to 0$, $H \to \max(0, F + G - 1)$ and as $\psi \to \infty$, $H \to \min(F, G)$. Thus this class contains the Fréchet boundary distributions and the distribution corresponding to independent r.v's. Mosteller (1968) derives this distribution by the process of standardizing a 2×2 table by row and column multiplications.

If F and G are absolutely continuous d.f's then the joint p.d.f. of X and Y exists, because by (8.1.4) H is differentiable with respect to F and G, and this implies that it is differentiable with respect to x and y. We have

$$h = fg\, \partial^2 H/\partial F\, \partial G$$

and, by (8.1.4), we obtain

$$h(x,y;\psi) = \frac{\psi fg\{(\psi-1)(F+G-2FG)+1\}}{\{S^2-4\psi(\psi-1)FG\}^{3/2}}, \quad \psi > 0, \tag{8.1.7}$$

where S is given by (8.1.5).

In view of the motivation given above to derive this class, the distributions it contains are described as contingency-type or, briefly, C-type. When X and Y are normal variables, the distribution given by (8.1.4) is a C-type normal distribution and when $F = x$ and $G = y$ in (8.1.4), we get the C-type uniform distribution. In fact, (8.1.6) is the p.d.f. of the C-type uniform distribution.

8.2 Properties

8.2a Distributional properties

In this section, we assume that $F(x)$ and $G(y)$ are absolutely continuous, so that F and G have inverses.

The conditional d.f. $K(y|x,\psi)$ of Y given X is equal to $\partial H/\partial G$, and, by (8.1.4), it is given by

$$K(y|x,\psi) = \frac{1}{2} + \frac{(\psi+1)G-(\psi-1)F-1}{2\{S^2-4\psi(\psi-1)FG\}^{1/2}}. \tag{8.2.1}$$

If we denote (8.2.1) by $K^*(y|F,\psi)$ then we have

$$K^*(y|F,\psi) = K^*(y|1-F,\ 1/\psi). \tag{8.2.2}$$

If the r.v's X and Y both have symmetric distributions then from (8.1.4) the r.v. (X,Y) has a symmetric distribution. Further, if (X,Y) has a symmetric distribution about (0,0) then

$$h(-x,y;\psi) = h(x,y;1/\psi). \tag{8.2.3}$$

Let $\mu_{rs}(\psi)$ denote the central moment of order (r,s). On using (8.2.3) and the following result true for any function $v(x,y)$

$$\int_{-\infty}^{\infty}\int_{-\infty}^{\infty} v(x,y)\,dx\,dy = \int_0^{\infty}\int_0^{\infty}\{v(x,y)+v(-x,-y)+v(x,-y)+v(-x,y)\}\,dx\,dy, \tag{8.2.4}$$

we have a result due to Steck (1968):

$$\begin{aligned} \mu_{rs}(\psi) &= \mu_{rs}(1/\psi) && \text{if } r,s \text{ are even,} \\ &= -\mu_{rs}(1/\psi) && \text{if } r,s \text{ are odd,} \\ &= 0 && \text{if either } r,s \text{ is odd.} \end{aligned}$$

In particular, if $\rho(\psi)$ denotes $\operatorname{corr}(X,Y)$ of this class, we have for this case

$$\rho(\psi) = -\rho(1/\psi). \tag{8.2.5}$$

Now, we show that for any distribution

$$\operatorname{corr}(X,Y) = \frac{1}{\sigma_1 \sigma_2} \int_{-\infty}^{\infty}\int_{-\infty}^{\infty} (H - FG)\,dx\,dy. \tag{8.2.6}$$

Using the result (8.2.4), we split the integral into four parts so that

$$\operatorname{cov}(X,Y) = I_1 + I_2 + I_3 + I_4, \tag{8.2.7}$$

where $$I_1 = \int_0^{\infty}\int_0^{\infty} xy\{h(x,y) - f(x)g(y)\}\,dx\,dy$$

and so on. We can write

$$I_1 = \int_0^{\infty}\int_0^{\infty}\int_0^{x}\int_0^{y} \{h(x,y) - f(x)g(y)\}\,du\,dv\,dx\,dy,$$

and if E(X), E(Y), and E(XY) exist, then the order of integration in I_1 can be changed by applying Fubini's theorem. On doing so, we obtain

$$I_1 = \int_0^{\infty}\int_0^{\infty} \left[\int_u^{\infty}\int_v^{\infty} \{h(x,y) - f(x)g(y)\}\,dx\,dy\right] du\,dv$$

or $$I_1 = \int_0^{\infty}\int_0^{\infty} (H - FG)\,dx\,dy.$$

Similar expressions can be derived for I_2, I_3 and I_4. On substituting these values for I_1, I_2, I_3 and I_4 in (8.2.7) and using (8.2.4), we get the desired result. It may be noted that (8.2.6) was given by Hoeffding (1940) whose proof is also given in Lehmann (1966). An extension of this result is given by Mardia (1967a).

We easily see from (8.2.6) that $\rho(1) = 0$ if X and Y are independent. If $\psi > 1$ then by (8.1.1), $H > FG$, so that from (8.2.6), $\rho(\psi) > 0$. Similarly if $\psi < 1$, we find that $\rho(\psi) < 0$. Now $\psi > 1$ implies that X and Y are positively associated, while $\psi < 1$ implies that X and Y are negatively associated (Kendall and Stuart, 1967, p. 538, equations (33.6) and (33.7)). Therefore, the correlation and contingency coefficients reflect the same trend in the relationship between X and Y. In addition, when $\psi = \infty$ we have $F(x) = G(y)$, so that if $F = G$ then $x = y$. Hence in this case $\psi = \infty$ implies $\rho(\psi) = 1$ and furthermore, if the r.v. X has symmetric distribution, from (8.2.5) it follows that $\rho(\psi) = -1$ for $\psi = 0$.

8.2b Relation to other families

With the help of (8.1.6), we can easily verify that the C-type uniform distribution is a solution of the Pearson differential equations given by (2.1.9).

We now establish its relation with another family. Let $\alpha = \psi - 1$, so that $\alpha = 0$ when X and Y are independent r.v's. We can write (8.1.4) as

$$H = \frac{1}{2}(F + G) + \frac{1}{2\alpha}[1 - \{1 + 2\alpha(F + G - 2FG) + \alpha^2(F - G)^2\}^{1/2}].$$

On using the binomial expansion, we immediately obtain

$$H = FG\{1 + \alpha(1 - F)(1 - G)\} + o(\alpha) \qquad (8.2.8)$$

which is the family introduced by Morgenstern (1956) and is studied in §9.2.

8.2c The inverse of K with applications

Let $K(y|x,\psi) = K$ where K is given. We show that

$$y = G^{-1}[\{B - (1 - 2K)D\}/(2A)], \qquad (8.2.9)$$

where $A = \psi + a(\psi - 1)^2$, $B = 2a\{F\psi^2 + (1 - F)\} + \psi(1 - 2a)$, (8.2.10)

$D = \psi^{1/2}\{\psi + 4aF(1 - F)(1 - \psi)^2\}^{1/2}$, $a = K(1 - K)$. (8.2.11)

From (8.2.1) we have

$$AG^2 - BG + C = 0, \qquad (8.2.12)$$

where $C = a\{1 + (\psi - 1)F\}^2$. We first note that $0 \leqslant a \leqslant \frac{1}{4}$, so that A, B and C are positive. Let G_1 and G_2 be the roots of (8.2.12) with

$$G_1 = (B - D|1 - 2K|)/(2A), \quad G_2 = (B + D|1 - 2K|)/(2A). \qquad (8.2.13)$$

If $K = \frac{1}{2}$ then these roots are equal to G_0 where

$$G_0 = \{1 + (\psi - 1)F\}/(\psi + 1). \qquad (8.2.14)$$

On substituting for G_0, G_1 and G_2 from (8.2.13) and (8.2.14), we find that

$$(G_0 - G_1)(G_2 - G_0) = \{C/(Aa)\}^{1/2}\,\psi(1 - 4a)\{F + \psi(1 - F)\}$$

which is positive since $0 \leqslant a \leqslant \frac{1}{4}$. Now $A \geqslant 0$, so that $G_1 \leqslant G_2$, and hence we have

$$G_1 \leqslant G_0 \leqslant G_2. \qquad (8.2.15)$$

From (8.2.1) and (8.2.14) we find that $-(G - G_0)(\psi + 1)/(1 - 2K)$

is the square root of a positive quantity, so that we must always have

$$(G - G_0)/(1 - 2K) \leqslant 0. \qquad (8.2.16)$$

On using (8.2.15) and (8.2.16) we see that if $K \leqslant \frac{1}{2}$ then $G = G_1$ and if $K \geqslant \frac{1}{2}$ then $G = G_2$, so that finally from (8.2.13) we have (8.2.9).

We now study the median regressions since $\mathrm{E}(Y|x)$ cannot be simplified in general. From (8.2.9) we have, on putting $K = \frac{1}{2}$, that the median regression of Y on X is given by

$$(\psi + 1)G(y) = 1 + (\psi - 1)F(x). \qquad (8.2.17)$$

Further, the scedastic curve of Y on X, based on the semi-interquartile range from (8.2.9), is

$$2y = G^{-1}[(2B^* + D^*)/(4A^*)] - G^{-1}[(2B^* - D^*)/(4A^*)], \qquad (8.2.18)$$

where A^*, B^* and D^* are the values of A, B and D at $a = 3/16$. In particular, for the C-type uniform distribution, the median regression of Y on X from (8.2.17) is linear and the scedastic curve of Y on X reduces to

$$y = 2[\psi\{4\psi + 3(\psi - 1)^2 x(1 - x)\}]^{1/2} / \{(3\psi + 1)(\psi + 3)\} \qquad (8.2.19)$$

which is an ellipse.

We now provide a simple method of drawing random samples from the C-type population. If we transform X and Y to U and V by Rosenblatt's transformation (1952):

$$u = F(x), \quad v = K(y|x),$$

then U and V are distributed independently as uniform random variables on (0,1). Hence a random pair (x,y) from this population is the solution of the equations $F = u$ and $K = v$, where (u,v) is a given random uniform pair. By taking $F = u$ and $K = v$ in (8.2.9) we obtain y, and x is evaluated from $F = u$. Hence the construction is reduced to the simpler problem of inverting F and G.

8.3 The C-type uniform distribution

In this section, we assume that F and G correspond to uniform distributions on (0,1). On writing $F = x$ and $G = y$ in (8.1.4), we get the d.f. of the C-type uniform distribution as

$$H_U(x,y;\psi) = \{1 + (x + y)(\psi - 1) - L^{1/2}\}/\{2(\psi - 1)\}, \quad (8.3.1)$$

where $$L = \{1 + (x + y)(\psi - 1)\}^2 - 4\psi(\psi - 1)xy. \quad (8.3.2)$$

The corresponding p.d.f. has already been given at (8.1.6), and the conditional d.f. of Y given X is

$$K_U(y|x,\psi) = \tfrac{1}{2} + \tfrac{1}{2}\{(\psi+1)y - (\psi-1)x - 1\}L^{-1/2}. \quad (8.3.3)$$

We have $$E(Y^r|x) = r\int_0^1 y^{r-1}\{1 - K_U(y|x,\psi)\}dy,$$

and substituting (8.3.3), this reduces to

$$E(Y^r|x) = \tfrac{1}{2} - \frac{r}{2}\int_0^1 y^{r-1}[\{(\psi+1)y - (\psi-1)x - 1\}L^{-1/2}]dy.$$

We can write $L = \{(\psi-1)y - (\psi+1)x + 1\}^2 + 4\psi x(1-x)$, and putting

$$(\psi-1)y - (\psi+1)x + 1 = b\sinh(\theta + c) \quad (8.3.4)$$

with $b = 2\surd\{\psi x(1-x)\}$, $c = \sinh^{-1}[\{1-(\psi+1)x\}/b]$

we find that

$$E(Y^r|x) = \tfrac{1}{2} - \frac{r}{2(\psi-1)^{r+1}}\{(\psi+1)I_r + (\psi-1)(1-x+x\psi)I_{r-1}\}, \quad (8.3.5)$$

where $$I_r = \int_0^{\log\psi}[\{1+x(\psi-1)\}\sinh\theta + \{1-x(\psi+1)\}(\cosh\theta - 1)]^r\,d\theta.$$

Hence $E(Y^r|x)$ is a polynomial of the rth degree in x. Furthermore, we have

$$E(Y|x) = \tfrac{1}{2} + \rho_U(x - \tfrac{1}{2}) \quad (8.3.6)$$

and $$\mathrm{var}(Y|x) = (3Q\rho_U - 2 - \rho_U^2)(x-\tfrac{1}{2})^2 + \tfrac{1}{4}(1 - Q\rho_U), \quad (8.3.7)$$

where $$\rho_U = \frac{\psi+1}{\psi-1} - \frac{2\psi}{(\psi-1)^2}\log\psi \quad (8.3.8)$$

and $Q = (\psi+1)/(\psi-1)$. Hence the mean regression of Y on X is linear and the scedastic curve is a parabola.

On multiplying (8.3.6) by x and taking the expectation, we find that $\mathrm{corr}(X,Y) = \rho_U$. On multiplying (8.3.6) and (8.3.7) by the appropriate power of $(x-\tfrac{1}{2})$ and taking the expectation, we similarly have

$$\mu_{13} = \mu_{31} = \rho_U/80, \quad \mu_{22} = (4Q\rho_U - 1)/240. \quad (8.3.9)$$

For this case, we have obtained the scedastic curves based on the semi-interquartile range at (8.2.19). However, it is possible to obtain the scedastic curves based on the mean deviation. From (8.2.17) we have

$$\tilde{y}_x = \{1 + (\psi-1)x\}/(\psi+1).$$

On using the result $$E|X - a| = \int_{-\infty}^{a} F(x)dx + \int_a^{\infty}\{1 - F(x)\}dx,$$

we find, with the help of the transformation (8.3.4), that

$$\mathrm{E}[\{|Y - \tilde{y}_x|\}|x]$$

$$= \frac{2\psi^{1/2}}{(\psi-1)^2}\left\{\alpha - \psi^{1/2} - \psi^{1/2}(1-2x)\log\frac{\alpha+\psi^{1/2}(1-2x)}{(1-x)(1+\psi)}\right\}, \qquad (8.3.10)$$

where $\alpha = \sqrt{\{\psi + x(1-x)(1-\psi)^2\}}$. It can be seen that the scedastic curves given by (8.2.19) and (8.3.10) are both symmetrical about $x = \frac{1}{2}$.

8.4 The C-type normal distribution

When X and Y are $N(0, 1)$, the distribution given by (8.1.4) will be described as a C-type bivariate normal distribution. Let us denote its d.f. by $C(x,y;\psi)$.

On using (8.1.4) in (8.2.6) and the result (8.2.4), we find that corr(X,Y) for this distribution is given by

$$\rho_N = \frac{1}{(\psi-1)}\int_0^\infty\int_0^\infty [\psi + 1 - A\{B(x), B(y)\} - A\{1 - B(x), B(y)\}]\,dx\,dy, \qquad (8.4.1)$$

where $B(.)$ is the d.f. of $N(0, 1)$ and

$$A(u,v) = [\{1 + (\psi-1)(u+v)\}^2 - 4\psi(\psi-1)uv]^{1/2}.$$

This cannot be further simplified, and Mardia (1967a) has tabulated ρ_N for various values of ψ. However, some approximations to (8.4.1) can be derived. Let $B(x,y;\rho)$ be the d.f. of $N(0,0,1,1,\rho)$. Following Plackett (1965), we can relate ρ and ψ by

$$C(0,0;\psi) = B(0,0;\rho). \qquad (8.4.2)$$

We have $\quad B(0,0;\rho) = \frac{1}{4} + \arcsin\rho/(2\pi)$

and from (8.1.4) $\quad C(0,0;\psi) = \sqrt{\psi}/2(1+\sqrt{\psi})$

so that (8.4.2) gives $\quad \rho = \cos\{\pi/(1+\sqrt{\psi})\}. \qquad (8.4.3)$

We denote this approximation to ρ_N by ρ_1.

Alternatively, following Mardia (1967a) we can relate ρ and ψ by equating corr (F,G) for $B(x,y;\rho)$ and $C(x,y;\psi)$. Let us denote these correlation coefficients by $\rho_B(F,G)$ and $\rho_C(F,G)$ respectively. Now, $\rho_B(F,G)$ is Spearman's correlation coefficient for the normal population, so that

$$\rho_B(F,G) = \frac{6}{\pi}\arcsin(\tfrac{1}{2}\rho),$$

and $\rho_C(F,G)$ is equal to ρ_U given by (8.3.8). On substituting these values in $\rho_B(F,G) = \rho_C(F,G)$, we find

$$\rho = 2\sin[\pi(\psi^2 - 1 - 2\psi\log\psi)/\{6(\psi - 1)^2\}]. \qquad (8.4.4)$$

We denote this approximation to ρ_N by ρ_2.

Table 1 of Mardia (1967a) gives the values of ρ_N, ρ_U, ρ_1 and ρ_2 for given ψ, and from this table it is seen that

$$|\rho_N - \rho_U| < |\rho_N - \rho_2| < |\rho_N - \rho_1|$$

for all ψ except at $\psi = 0, 1, \infty$, where the correlation coefficients are equal. Thus ρ_U provides a better approximation to ρ_N than ρ_1 and ρ_2.

8.5 Estimation of ψ

Let (X_i, Y_i), $i = 1, \ldots, n$, be a random sample from the C-type population. We assume that the marginal distributions have known forms, the parameters of which can be estimated from (X_i) and (Y_i), $i = 1, \ldots, n$, separately. We deal here with the estimation of ψ.

8.5a The maximum likelihood estimator

Since the p.d.f. of the distribution is given explicitly by (8.1.7), we can write down the maximum likelihood equation for ψ, but the equation is complicated and the solution can be obtained only by a numerical process. However, for small values of $\psi - 1$, from (8.2.8), we have

$$\frac{\partial \log L}{\partial \psi} \doteq \sum_{i=1}^{n} (1 - 2F_i)(1 - 2G_i) - (\psi - 1)\sum_{i=1}^{n} (1 - 2F_i)^2 (1 - 2G_i)^2 .$$

Hence,
$$\hat{\psi} \doteq 1 + \sum_{i=1}^{n} (1 - 2F_i)(1 - 2G_i) \Big/ \sum_{i=1}^{n} (1 - 2F_i)^2 (1 - 2G_i)^2 \qquad (8.5.1)$$

and for large n,

$$\lim_{\psi \to 1} \operatorname{var}(\hat{\psi}) = \lim_{\psi \to 1} \left\{ \mathrm{E}\left(- \frac{\partial^2 \log L}{\partial \psi^2} \right) \right\}^{-1}$$

$$= \{n\mathrm{E}\,(1 - 2F)^2 (1 - 2G)^2\}^{-1},$$

so that
$$\lim_{\psi \to 1} \operatorname{var}(\hat{\psi}) \simeq 9/n. \qquad (8.5.2)$$

8.5b The cross-ratio

We now show that a simple estimator of ψ can be constructed. Suppose the joint distribution is divided into four quadrants by lines $X = x$ and $Y = y$. Let a, b, c and d be the frequencies of pairs (x_i, y_i) in the quadrants $(X \leqslant x, Y \leqslant y)$, $(X \leqslant x, Y > y)$, $(X > x, Y \leqslant y)$ and $(X > x, Y > y)$ respectively. A natural estimate of ψ is given by

$$\psi^{+} = ad/bc. \tag{8.5.3}$$

Now the frequencies a, b, c and d depend on the point of dichotomy (x,y), so that ψ^{+} can be used in practice after selecting (x,y). An optimum choice for (x,y) is the one which minimizes $\mathrm{var}(\psi^{+})$. From Kendall and Stuart (1967, p. 540), we have

$$\mathrm{var}(\psi^{+}) \simeq \frac{\psi^{2}}{n}(p_1^{-1} + p_2^{-1} + p_3^{-1} + p_4^{-1}), \tag{8.5.4}$$

where p_1, p_2, p_3 and p_4 are defined in § 8.1, so that

$$\mathrm{var}(\psi^{+}) \simeq \frac{\psi^{2}}{n}\left\{\frac{1}{H(x,y)} + \frac{1}{F(x) - H(x,y)} + \frac{1}{G(x) - H(x,y)} + \frac{1}{1 - F(x) - G(y) + H(x,y)}\right\}.$$

We now show that var (ψ^{+}) is minimized with respect to x and y if (x,y) is the population median vector, i.e. when $F(x) = G(y) = \frac{1}{2}$.

This problem of minimization is the same as minimizing the function

$$T(F,G) = \frac{1}{H} + \frac{1}{F - H} + \frac{1}{G - H} + \frac{1}{1 - F - G + H}$$

with respect to F and G. We show that $T(F,G)$ has one and only one minimum and this occurs at the point $F = G = \frac{1}{2}$. With the help of (8.1.2), we find that

$$T(F,G) = \frac{(1-\psi)\{1 + (1-\psi)(2H - F - G)\}}{\psi(FG - H)}, \quad \psi \neq 1,$$

$$= \frac{1}{F(1-F)G(1-G)}, \quad \psi = 1.$$

The problem for $\psi = 1$ is straightforward. For $\psi \neq 1$, we find, after using (8.1.2) and some simplification, that

$$\frac{\partial T}{\partial F} = \frac{(1-\psi)^2 G(1-G)\{\psi(F-G) + F + G - 1\}}{\psi(H - FG)^2\sqrt{S}}.$$

By interchanging F and G in the right-hand side we obtain $\partial T/\partial G$. On equating these derivatives to zero, we immediately find that the point $F = G = \frac{1}{2}$ is the only admissible turning point of $T(F,G)$. Now at $F = G = \frac{1}{2}$, we have

$$\frac{\partial^2 T}{\partial F^2} = \frac{\partial^2 T}{\partial G^2} = \frac{4(1+\sqrt{\psi})^2}{\psi^{3/2}}(1 + 2\sqrt{\psi} + 2\psi + \psi^2 + 2\psi\sqrt{\psi})$$

and
$$\frac{\partial^2 T}{\partial F\,\partial G} = \frac{4(1-\psi)(1+\sqrt{\psi})^4}{\psi^{3/2}}.$$

Therefore, it can be seen that at the point $(\frac{1}{2}, \frac{1}{2})$,

$$\frac{\partial^2 T}{\partial F^2} > 0, \quad \left(\frac{\partial^2 T}{\partial F^2}\right)\left(\frac{\partial^2 T}{\partial G^2}\right) > \left(\frac{\partial^2 T}{\partial F\,\partial G}\right),$$

so that this point is the minimum point of $T(F,G)$. Hence the optimum choice of (x,y) is the population median vector, and we now take this as our point of dichotomy. In practice, then, (x,y) is estimated by the sample median vector. Further, for $F = G = \frac{1}{2}$, we have from (8.1.4)

$$H = \sqrt{\psi}/2(1+\sqrt{\psi}),$$

so that in this case

$$p_1 = p_4 = \sqrt{\psi}/2(1+\sqrt{\psi}), \quad p_2 = p_3 = 1/2(1+\sqrt{\psi}).$$

Hence (8.5.4) becomes

$$\operatorname{var}(\psi^+) \simeq \frac{4}{n}\psi^{3/2}(1+\sqrt{\psi})^2. \tag{8.5.5}$$

From (8.5.2) and (8.5.5) we see that as $\psi \to 1$ the asymptotic efficiency of ψ^+ relative to $\hat{\psi}$, tends to $9/16 = 0{\cdot}56$. Hence the estimator ψ^+ is inefficient. On the other hand, the ease of computation of ψ^+ as compared to $\hat{\psi}$ is an advantage.

8.5c An efficient estimator

We now proceed to give an asymptotically efficient estimator of ψ. Since the forms of F and G are known, we can calculate the sample value of corr(F,G). We denote it by $\bar{r}$ and define a new estimate of ψ, say ψ^*, as a solution of

$$\bar{r} = \rho_U, \tag{8.5.6}$$

where ρ_U is the population value of corr(F,G) and is given by (8.3.8). From Kendall and Stuart (1969, p. 236) for large samples we have

$$\operatorname{var}(\bar{r}) \simeq \frac{1}{4n}\{\rho^2(\gamma_{40}+\gamma_{04}+2\gamma_{22}) - 4\rho(\gamma_{31}+\gamma_{13}) + 4\gamma_{22}\},$$

where $\gamma_{ij} = \mu_{ij}/\sigma_1^i\sigma_2^j$. On using the moments given in (8.3.9), we find after simplification that

$$\operatorname{var}(\bar{r}) \simeq 3A(\psi)/\{5n(\psi-1)\}, \tag{8.5.7}$$

where $A(\psi) = \psi(2\rho_U - 1)(\rho_U - 1)^2 + (2\rho_U + 1)(\rho_U + 1)^2. \quad (8.5.8)$

Using the asymptotic result

$$\operatorname{var}(\psi^*) \simeq \left(\frac{d\psi^*}{d\bar{r}}\right)^2 \operatorname{var}(\bar{r}) \qquad (8.5.9)$$

and (8.5.6), we have

$$\operatorname{var}(\psi^*) \simeq \frac{3\psi^2(\psi - 1)A(\psi)}{5n\{(\psi - 1) - (\psi + 1)\rho_U\}^2}. \qquad (8.5.10)$$

Thus ψ^* is a consistent estimator of ψ. As $\psi \to 1$, from (8.5.10), we find that $\operatorname{var}(\psi^*) \to 9/n$ and so by (8.5.2), ψ^* is an asymptotically equivalent estimator to $\hat{\psi}$. Consequently ψ^* is an asymptotically efficient estimator of ψ. The asymptotic efficiency, $e(\psi^+, \psi^*)$, of ψ^+ relative to ψ^* can be obtained from (8.5.5) and (8.5.10). It has been shown in Mardia (1967a) that in the region of greatest practical interest $|\rho_U| < 0{\cdot}99$, the asymptotic efficiency $e(\psi^+, \psi^*)$ lies between 0·46 and 0·56.

Another possible estimator of ψ may be obtained by equating the sample correlation coefficient r with the population value. Mardia (1967a) has shown that as $\psi \to 1$, the asymptotic efficiency of this estimator relative to ψ^* is less than or equal to 1, and the equality is achieved only for the C-type uniform population. Therefore the estimator is not recommended.

8.6 Methods of fitting

An efficient method of fitting a C-type distribution is as follows. (i) Fit each of the marginal distributions by a curve from a standard univariate family. (ii) Calculate $\bar{r} = \operatorname{corr}\{F(x_i), G(y_i)\}$ using F and and G obtained in (i), and obtain ψ^* from (8.5.6) or using Table 4 of Mardia (1967a).

An alternative method which is rather rough-and-ready is as follows. (a) Fit each marginal distribution by a curve from Johnson's family (Johnson, 1949a) where the values of $F(x)$ may be computed by using the logistic approximation (4.3.18). (b) Compute the estimate ψ^+ defined by (8.5.3).

We can calculate the expected frequency in the cell

$$(x_1 = x - \tfrac{1}{2}h \leqslant X \leqslant x + \tfrac{1}{2}h = x_2, \quad y_1 = y - \tfrac{1}{2}k \leqslant Y \leqslant y + \tfrac{1}{2}k = y_2)$$

from

$$\frac{N}{2(\psi - 1)}\{A(x_2, y_1) + A(x_1, y_2) - A(x_1, y_1) - A(x_2, y_2)\}, \qquad (8.6.1)$$

where N denotes the number of observations and

$$A(x,y) = \sqrt{\{S^2 - 4\psi(\psi - 1)F(x)G(y)\}}.$$

The mean regression and scedastic curves of $F(X)$ on $G(Y)$ and of $G(Y)$ on $F(X)$ can be plotted using the results of (i) and (ii) in (8.3.6) and (8.3.7). Higher array curves can be plotted by using (8.3.5). The median regression and scedastic curves given by (8.2.17) and (8.2.18) can be plotted with the help of (a) and (b).

Example 8.1. We now illustrate the first method of fitting by again using Johanssen's data of beans given in Table 3.1.

From § 4.2e, we have

$$F(x) = B\left\{2{\cdot}38 + 2{\cdot}64 \sinh^{-1}\left(\frac{x - 16{\cdot}0745}{1{\cdot}5192}\right)\right\} \tag{8.6.2}$$

and

$$G(y) = B\left\{2{\cdot}13 + 3{\cdot}55 \sinh^{-1}\left(\frac{y - 8{\cdot}6195}{0{\cdot}9721}\right)\right\}. \tag{8.6.3}$$

Using (8.6.2) and (8.6.3), it is found that $\bar{r} = 0{\cdot}7233$, and on interpolating in Table 4 of Mardia (1967a) we obtain

$$\hat{\psi} = 14{\cdot}560.$$

The mean regression and scedastic curves of $G(Y)$ on $F(X)$ and of $F(X)$ on $G(Y)$ are shown in Fig. 8.1–8.4. The plotted circles show the corresponding values calculated from the observed array distributions. From Fig. 8.1 and 8.3 we observe that the regression lines are close fits. From Fig. 8.2 and 8.4 we see that the general trend and order of the observed points are clearly parabolic but that the fitted curves, although parabolic, do not give a close fit. However, the number of observations in the tails of the two marginal distributions are very small, so that the standard errors of the observed array variances corresponding to the tails are appreciable. In § 3.4a and § 4.2e we have seen that in fitting the Type AA and S_{IJ} surfaces the scedastic curves are again not well reproduced. However, in this case, the trend of the observed values can easily be recognized and the fitted curves are free from the arbitrary oscillation which is found in fitting Type AA surfaces (§ 3.4a).

In this case, contrary to the S_{IJ} surfaces and Type AA surfaces, the expected frequencies can easily be computed. Table 8.1 gives the observed and the expected frequencies for the data. A comparison of the frequencies indicates that the fit, while not being good, does seem reasonable. Table 8.2 is an abridged form of Table 8.1 obtained by wider grouping to carry out a χ^2 test of the goodness of fit. The

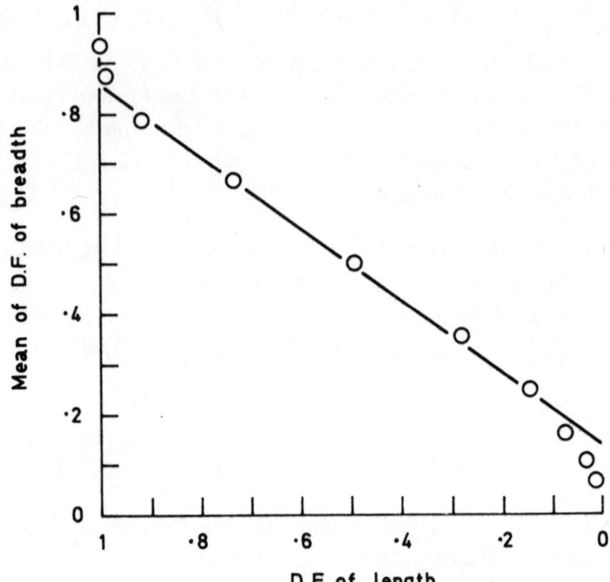

Fig. 8.1 Regression curve of the d.f. of the breadth of beans on the d.f. of the length of beans (C-type distribution)

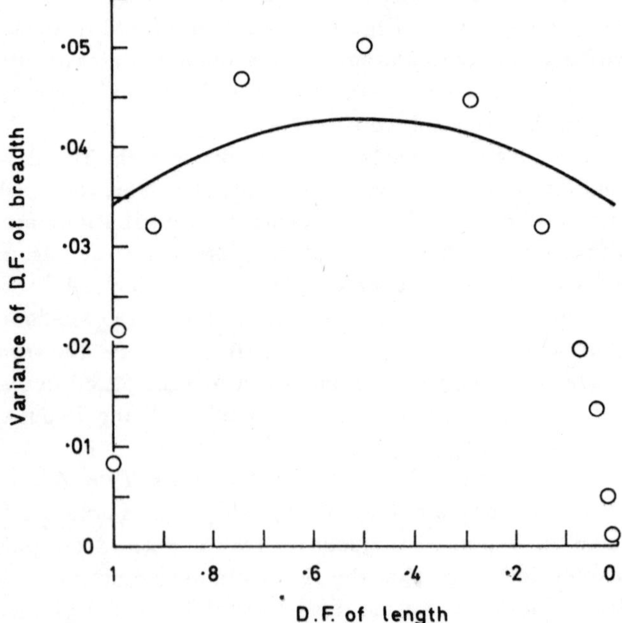

Fig. 8.2 Scedastic curve of the d.f. of the breadth of beans on the d.f. of the length of beans (C-type distribution)

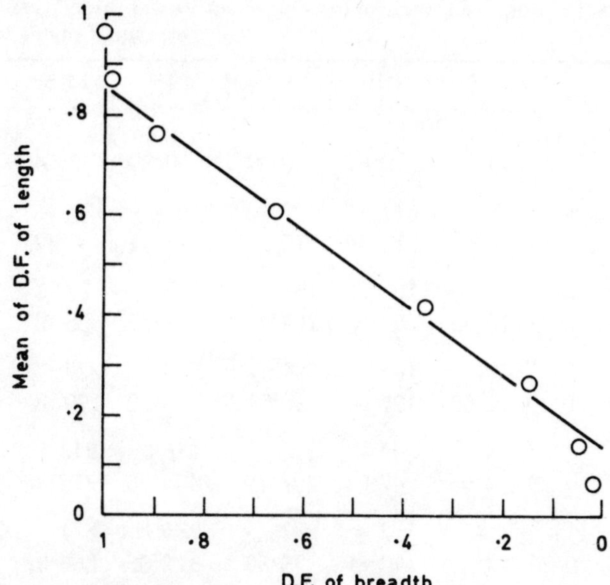

Fig. 8.3 Regression curve of the d.f. of the length of beans on the d.f. of the breadth of beans (C-type distribution)

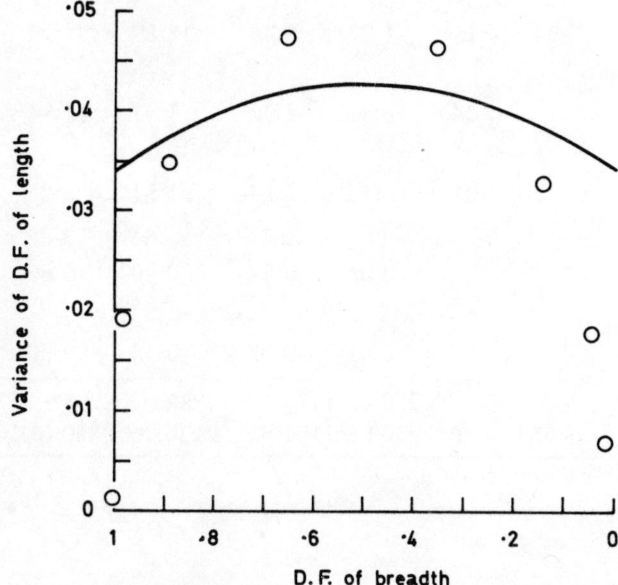

Fig. 8.4 Scedastic curve of the d.f. of the length of beans on the d.f. of the breadth of beans (C-type distribution)

Table 8.1 Observed frequencies (*upper entry*) and the corresponding expected measured in millimetres (Johans-

$Y \downarrow$	$X \rightarrow$ 17	16·5	16	15·5	15	14·5	14	13·5
9·125	–	2	–	–	3	–	–	–
	0·01	0·11	0·67	0·90	0·39	0·14	0·05	0·02
8·875	4	8	17	19	–	–	–	–
	0·12	1·83	11·64	16·17	7·12	2·47	0·95	0·40
8·625	2	23	101	156	93	23	2	–
	0·71	10·95	84·95	164·44	78·77	25·67	9·49	3·94
8·375	–	18	105	494	574	227	56	9
	0·79	12·63	125·94	558·64	573·22	179·80	55·64	21·15
8·125	–	4	44	375	956	913	362	73
	0·29	4·66	47·01	307·49	1063·36	912·92	280·47	85·29
7·875	–	–	7	81	385	871	794	330
	0·09	1·42	13·58	73·87	312·65	866·08	775·86	297·78
7·625	–	–	1	4	65	236	469	361
	0·03	0·46	4·29	20·96	68·37	193·15	399·05	380·36
7·375	–	–	–	–	6	23	91	137
	0·01	0·15	1·37	6·44	18·82	44·21	90·19	135·66
7·125	–	–	–	–	–	1	13	18
		0·04	0·41	1·92	5·41	11·84	22·31	34·47
6·875		–	–	–	–	–	–	1
		0·01	0·12	0·54	1·51	3·23	5·90	9·00
6·625			–	–	–	–	–	–
			0·03	0·14	0·40	0·86	1·56	2·36
6·375			–	–	–	–	–	–
			0·01	0·05	0·14	0·30	0·55	0·83
Totals	6	55	275	1129	2082	2294	1787	929
	2·05	32·26	290·02	1151·56	2130·16	2240·67	1642·02	971·26

frequencies (*lower entry*) of the number of beans of length X and breadth Y
sen's data, cited by Pretorius (1930))

13	12·5	12	11·5	11	10·5	10	9·5	Totals
–								5
0·01								2·30
–	–	–	–	–				48
0·18	0·08	0·04	0·02	0·01				41·03
–	–	–	–	–	–	–	–	400
1·73	0·78	0·36	0·16	0·08	0·04	0·02	0·02	382·11
–	–	–	–	–	–	–	–	1483
8·95	3·96	1·79	0·82	0·38	0·18	0·08	0·08	1544·05
12	3	–	–	–	–	–	–	2742
31·91	13·34	5·87	2·65	1·22	0·57	0·27	0·25	2757·57
89	19	3	–	–	–	–	–	2579
99·47	37·71	15·75	6·94	3·15	1·46	0·69	0·65	2507·15
175	55	27	4	–	–	–	–	1397
190·54	77·94	32·34	14·06	6·33	2·91	1·37	1·29	1393·45
124	78	37	22	11	–	1	–	530
122·70	72·49	35·36	16·39	7·57	3·53	1·67	1·57	558·13
28	35	25	32	11	6	1	–	170
38·80	30·44	18·22	9·45	4·62	2·21	1·06	1·00	182·20
9	8	21	12	13	7	1	–	72
10·56	9·12	6·04	3·36	1·71	0·83	0·40	0·38	52·71
–	–	2	–	1	4	3	–	10
2·79	2·48	1·69	0·96	0·50	0·24	0·12	0·11	14·24
–	1	–	–	–	1	1	1	4
0·98	0·88	0·61	0·35	0·18	0·09	0·04	0·04	5·05
437	199	115	70	36	18	7	1	9440
508·62	249·22	118·07	55·16	25·75	12·06	5·72	5·39	9439·99

Table 8.2 The observed and expected frequencies of the number of beans of length X and breadth Y obtained from Table 8.1 by wider grouping

Y ↓	$X \rightarrow$ 16	15·5	15	14·5	14	13·5	13	12·5	12
8·625	280 250·35	175 181·51				121 132·94			
8·375		494 558·64	574 573·22			292 272·83			
8·125	48 51·96	375 307·49	956 1063·36	913 912·92			450 421·84		
7·875		88 88·96	385 312·65	871 866·08	794 775·86			441 463·60	
7·625					469 399·05	361 380·36		261 326·78	
7·375					91 90·19	137 135·66	124 122·70	78 72·49	71 66·09
7·125			336 385·22				169 162·58		
6·875							86 64·66		

Total observed frequencies = 9440
Total expected frequencies = 9439·99
Value of χ^2 = 101·1

grouping is rather favourable to the null hypothesis. The expected frequencies for each group are not less than 50 and the value of $\chi^2 = 101{\cdot}1$. There are 27 groups, and 9 parameters are estimated. Now the 1 per cent value of χ^2 with 17 d. of fr. is 33·41, and hence the fit is not good. The observed value of χ^2 can be compared with Johnson's value of $\chi^2 = 87{\cdot}1$ with 9 d. of fr. on fitting the S_U curve to the length and $\chi^2 = 17{\cdot}5$ with 5 d. of fr. fitting the S_U curve to the breadth.

8.7 Other applications

It has been shown in Plackett (1965) and Mardia (1967a) that the C-type normal distribution provides some approximations to the d.f. of the standard bivariate normal. Further, these approximations are used to obtain simple methods of estimating the correlation coefficient of an $r \times c$ contingency table. Mardia (1967a) has examined the adequacy of these methods in computing the tetrachoric correlation coefficient. The asymptotic performance of various tests of independence for this kind of alternative has been investigated by Mardia (1969).

CHAPTER 9

TRIVARIATE REDUCTION AND OTHER METHODS

9.1 Trivariate reduction

W.F.R. Weldon, in his well-known dice problem, first constructed a bivariate binomial distribution by taking

$$X = X_1 + X_2, \quad Y = X_1 + X_3, \tag{9.1.1}$$

where X_1, X_2 and X_3 are independent binomial variables. This approach has recently been generalized by Arnold (1967). Let X_i, $i = 1,2,3$ be three independent variables with corresponding d.f's $F_0(x_i; \lambda_i)$, $i = 1,2,3$, where the λ's are parameters. Suppose there exists a function T such that the d.f. of

$$X = T(X_1, X_2) \tag{9.1.2}$$

is

$$F(x) = F_0(x; \lambda_1 + \lambda_2). \tag{9.1.3}$$

Further, this implies that the d.f. of

$$Y = T(X_1, X_3) \tag{9.1.4}$$

is

$$G(y) = F_0(y; \lambda_1 + \lambda_3). \tag{9.1.5}$$

Thus the r.v. (X,Y) defined by (9.1.2) and (9.1.4) has a bivariate $F_0(x; \lambda)$ distribution.

This method can be described as "trivariate reduction", in contrast to the translation method based on two independent r.v's (see Chapter 4). We now consider the existence of $F_0(x; \lambda)$ for some simple T-transformations and give important members of the resulting families.

Case I. Suppose

$$T(X_1, X_2) = X_1 + X_2, \tag{9.1.6}$$

that is, X and Y are given by (9.1.1). Now (9.1.3) implies that the d.f. $F_0(x; \lambda)$ possesses the additive property, so that if $\phi(t, \lambda)$ is the characteristic function for $F_0(x; \lambda)$ then

$$\phi(t, \lambda_1 + \lambda_2) = \phi(t, \lambda_1)\, \phi(t, \lambda_2). \tag{9.1.7}$$

This is a familiar functional equation in probability theory (see Feller, 1968, pp. 459–60) for $\lambda > 0$, and it leads to

$$\phi(t, \lambda) = [\phi(t)]^{\lambda} \tag{9.1.8}$$

for some characteristic function $\phi(t)$. Hence, distributions satisfying (9.1.8) will belong to this class. Before we give any members, it is important to note that

$$\rho = \operatorname{corr}(X, Y) = \frac{\bar{\sigma}_1^2}{\{(\bar{\sigma}_1^2 + \bar{\sigma}_2^2)(\bar{\sigma}_1^2 + \bar{\sigma}_3^2)\}^{1/2}}, \tag{9.1.9}$$

where $\bar{\sigma}_i^2 = \operatorname{var}(X_i)$, $i = 1,2,3$. As with the bivariate normal, $\rho = 0$ if and only if X and Y are independent, and $\rho = 1$ if and only if X and Y are linearly dependent. However, here $\rho \geqslant 0$. In fact, if $F_0(x;\lambda)$ is $N(0,\lambda)$ then from (9.1.8) we find that (X, Y) is bivariate normal, although $\rho \geqslant 0$. Similarly, when $F_0(x;\lambda)$ is the d.f. of the gamma distribution with index parameter λ then (X, Y) has Cherian's bivariate gamma distribution (§ 10.10), and when $F_0(x;\lambda)$ is Poisson with parameter λ then (X, Y) has the bivariate Poisson (§ 10.3).

Eagleson (1964) obtains a canonical representation (§ 7.3) of this class in an important particular case. Let $\{x_i(X_0)\}$ be a complete set of orthogonal polynomials defined on $F_0(x)$. Suppose that the generating function for $\{x_i(X_0)\}$ is of the form

$$\sum_{i=0}^{\infty} \frac{x_i(x_0)t^i}{i!} = v(t)\, e^{x_0 u(t)},$$

where $u(t)$ and $v(t)$ are real power series in t and $v(0) = 1$, $u(0) = 0$ and $u'(0) = 1$. For this case, he proves that the pairs of the orthonormal polynomials defined on the marginal distributions are canonical variables and that the ith canonical correlation is given by

$$\rho_i = \frac{C_i(\lambda_1)}{[C_i(\lambda_1+\lambda_2)\, C_i(\lambda_1+\lambda_3)]^{1/2}},$$

where

$$C_i(\lambda) = \int_{-\infty}^{\infty} x_i^2(x)\, dF_0(x,\lambda), \quad i = 1,2,.., \infty,$$

so that ρ_i depends only on the normalizing factor of the ith orthogonal polynomial $x_i(X_0)$. From this result he derives the canonical representation of the bivariate normal (§ 10.6), of Cherian's bivariate gamma (§ 10.10), of the bivariate Poisson (§ 10.3), and of some other distributions. The reader is referred to Eagleson (1964) for further details.

Case II. Suppose that

$$T(X_1, X_2) = \min(X_1, X_2). \tag{9.1.10}$$

In this case, (9.1.3) is true if

$$P\{X = \min(X_1, X_2) > x\} = P(X_1 > x)\, P(X_2 > x).$$

Hence, if $\bar{F}_0(x; \lambda) = 1 - F_0(x; \lambda)$ for $\lambda > 0$, then

$$\bar{F}_0(x; \lambda_1)\, \bar{F}_0(x; \lambda_2) = \bar{F}_0(x; \lambda_1 + \lambda_2)$$

which is of the form (9.1.7) and leads to

$$\bar{F}_0(x; \lambda) = \{\bar{F}_0(x)\}^{\lambda}$$

for some d.f. $F_0(x) = 1 - \bar{F}_0(x)$. It can now be easily verified that Marshall and Olkin's bivariate exponential distribution (§ 10.11) belongs to this class with $\bar{F}_0(x) = e^{-x}$. One can similarly construct bivariate Pareto, Weibull and geometric distributions.

The case of $T(x,y) = \max(x,y)$ can also be treated in this way.

Bhuchongkul (1964) has studied the Pitman efficiency of some tests of independence for alternatives of the form

$$X = \theta X_1 + (1 - \theta) X_2, \quad Y = \theta X_1 + (1 - \theta) X_3,$$

where $0 \leqslant \theta \leqslant 1$.

9.2 One-parameter families

In Chapters 4 and 8 we have studied one-parameter families with specified margins and which contain the three members

$$H_{-1}(x,y) = \max(F + G - 1, 0),\ H_0(x,y) = FG,\ H_1(x,y) = \max(F,G).$$

There are also other interesting families which we briefly survey, but they do not all contain these three distributions.

Morgenstern (1956) introduced the family with d.f.

$$H(x,y) = F(x)\, G(y)\, [1 + \alpha\{1 - F(x)\}\{1 - G(y)\}], \quad -1 \leqslant \alpha \leqslant 1. \tag{9.2.1}$$

For the continuous r.v. (X,Y), the p.d.f. is given by

$$h(x,y) = f(x) g(y)[1 + \alpha\{2F(x) - 1\}\{2G(y) - 1\}], \quad -1 \leqslant \alpha \leqslant 1. \tag{9.2.2}$$

Hence, if $F(x)$ and $G(y)$ correspond to uniform distributions on (0,1), the p.d.f. of Morgenstern's uniform distribution is given by

$$h(x,y) = 1 + \alpha(2x - 1)(2y - 1), \quad 0 < x,y \leqslant 1, \quad -1 \leqslant \alpha \leqslant 1. \tag{9.2.3}$$

We have in this case

$$E(Y \mid x) = \tfrac{1}{2} + \alpha(2x-1)/6, \quad \text{var}(Y \mid x) = \{3 - \alpha^2(2x-1)^2\}/36$$

and
$$\rho = \alpha/3.$$

Gumbel (1958) noted that (9.2.2) differed from the standard bivariate normal even when $F(x) = B(x)$ and $G(y) = B(y)$, where $B(.)$ is the d.f. of $N(0,1)$. Gumbel (1960) and Gumbel (1961) examined the bivariate exponential and the bivariate logistic members, respectively. Further, we have seen in Chapter 8 that the contingency-type distributions are of this form for ψ near 1.

Farlie (1960) extended (9.2.1) to the form

$$H(x,y) = F(x)\,G(y)\{1 + \alpha A(F)B(G)\}, \qquad (9.2.4)$$

where $A(F)$ and $B(G)$ are chosen so that (9.2.4) is a d.f. In fact, $H(x,y)$ is a d.f. if

(i) $A(F) \to 0$ as $F \to 1$, $B(G) \to 0$ as $G \to 1$,

(ii) $A(F)$ and $B(G)$ have bounded first differential coefficients with respect to their arguments, and

(iii)
$$-\frac{1}{M'} \leqslant \alpha \leqslant \frac{1}{M''},$$

where $M' = \max(M_2M_3, M_1M_4)$ and $M'' = \max(M_1M_3, M_2M_4)$ with

$$-M_1 \leqslant \{\partial(FA)/\partial F\} \leqslant M_2 \text{ and } -M_3 \leqslant \{\partial(GB)/\partial G\} \leqslant M_4.$$

The conditions (ii) and (iii) are sufficient to make $H(x,y)$ a monotonic function of x and y, i.e.

$$\frac{\partial^2 H}{\partial x\,\partial y} = \frac{\partial F}{\partial x}\frac{\partial G}{\partial y}\left\{1 + \alpha\,\frac{\partial(FA)}{\partial F}\,\frac{\partial(GB)}{\partial G}\right\} \geqslant 0.$$

It can be seen that the regression of Y on X is

$$E(Y \mid x) = \nu + \alpha C\left\{\frac{\partial(FA)}{\partial F}\right\},$$

where $C = \int_{-\infty}^{\infty} y\{\partial(GB)/\partial y\}\,dy$ and $\nu = E(Y)$.

Farlie (1960) has considered various special cases. The p.d.f. corresponding to (9.2.4) is

$$dH(x,y) = f(x)g(y)\left\{1 + \alpha\,\frac{d(FA)}{dF}\,\frac{d(GB)}{dG}\right\}. \qquad (9.2.5)$$

It should be noted that neither $H_{-1}(x,y)$ nor $H_1(x,y)$ is a member of (9.2.4). Farlie (1960, 1961) has examined the asymptotic performance of various tests of independence for this kind of alternative.

Gumbel (1961) has introduced the class

$$\{-\log H(x,y)\}^m = \{-\log F(x)\}^m + \{-\log G(y)\}^m, \quad m \geqslant 1, \qquad (9.2.6)$$

and has examined the bivariate logistic distribution belonging to this class. Gumbel and Mustafi (1967) have examined the bivariate distribution obtained from (9.2.6) by choosing

$$F(x) = \exp(-e^{-x}), \quad G(y) = \exp(-e^{-y}), \qquad (9.2.7)$$

so that the margins follow the first asymptotic distribution of the largest values. From (9.2.6), as $m \to \infty$, we find that $H(x,y) \to H_1(x,y)$ and that for $m = 1$, $H(x,y) = H_0(x,y)$, but as $m \to 0$, $H(x,y) \to 0$ for any fixed values of x and y.

Sibuya (1960) developed the family

$$\log H(x,y) = \log H_0(x,y) - a\,\frac{\{\log F(x)\log G(y)\}}{\log H_0(x,y)}, \quad 0 \leqslant a \leqslant 1, \qquad (9.2.8)$$

as a stable asymptotic distribution of largest values from given initial distributions. Gumbel and Mustafi (1967) have studied (9.2.8) for F and G defined by (9.2.7), and a criterion to distinguish between (9.2.6) and (9.2.8) for this case has also been given. For $a = 0$, $H(x,y) = H_0(x,y)$, but (9.2.8) does not contain $H_{-1}(x,y)$ and $H_1(x,y)$.

9.3 Bivariate exponential-type distributions

From Darmois–Koopman–Pitman distributions, we have

$$dH(x,y) = a(x,y)\exp\{x\theta_1 + y\theta_2 - q(\theta_1, \theta_2)\}, \qquad (9.3.1)$$

where θ_1 and θ_2 are two parameters. This is described as the class of bivariate exponential-type distributions and has been studied by Bildikar and Patil (1968). The marginal and conditional distributions are again of the exponential type. In subsequent discussions, we shall assume for simplicity that the random variable (X,Y) is continuous, although the conclusions are applicable in the general case. The moment generating function of (9.3.1) is

$$E(e^{t_1X+t_2Y}) = e^{-q(\theta_1,\theta_2)} \int_{-\infty}^{\infty}\int_{-\infty}^{\infty} a(x,y)\exp\{x(\theta_1+t_1) + y(\theta_2+t_2)\}\,dx\,dy.$$

Now,
$$\int_{-\infty}^{\infty}\int_{-\infty}^{\infty} a(x,y)\exp(x\theta_1 + y\theta_2)\,dx\,dy = e^{q(\theta_1,\theta_2)},$$
so that the moment generating function of (X,Y) is
$$\exp\{q(\theta_1 + t_1, \theta_2 + t_2) - q(\theta_1, \theta_2)\}. \tag{9.3.2}$$

Consequently, the cumulant generating function of (X,Y) is
$$C(t_1, t_2) = q(\theta_1 + t_1, \theta_2 + t_2) - q(\theta_1, \theta_2). \tag{9.3.3}$$

Let κ_{ij} be the cumulant of order (i,j). We have
$$\begin{aligned}\kappa_{i+1,j} &= \left[\frac{\partial^{i+1}}{\partial t_1^{i+1}}\frac{\partial^j}{\partial t_2^j} C(t_1, t_2)\right]_{t_1 = t_2 = 0} \\ &= \left[\frac{\partial^i}{\partial t_1^i}\frac{\partial^j}{\partial t_2^j}\left\{\frac{\partial}{\partial t_1} q(\theta_1 + t_1, \theta_2 + t_2)\right\}\right]_{t_1 = t_2 = 0} \\ &= \frac{\partial^{i+1}}{\partial \theta_1^{i+1}}\frac{\partial^j}{\partial \theta_2^j} q(\theta_1, \theta_2) \quad \text{if } i > 0.\end{aligned}$$

Hence it follows that
$$\kappa_{i+1,j} = \frac{\partial \kappa_{ij}}{\partial \theta_1}, \quad \kappa_{i,j+1} = \frac{\partial \kappa_{ij}}{\partial \theta_2} \tag{9.3.4}$$
for all $i, j \geqslant 0$ such that $i + j \geqslant 1$. Further, if $k_{11} = 0$ then (9.3.4) implies that $\kappa_{ij} = 0$ for all $i, j \geqslant 1$. Consequently, X and Y are independent if and only if they are uncorrelated.

It should be noted that (9.3.1) includes the bivariate binomial distribution (§ 10.1), the bivariate negative binomial distribution (§ 10.4), the bivariate logarithmic series (§ 10.5), and the bivariate normal distribution (§ 10.6).

In the discrete case, (9.3.1) can be written in the form
$$p(x,y) = a(x,y)\,\theta_1^x\,\theta_2^y / b(\theta_1, \theta_2),$$
which is the bivariate generalized power-series distribution studied by Khatri (1959) and Patil (1965).

9.4 Mixtures

If h_1 and h_2 are given bivariate p.d.f's both having the marginal p.d.f's f and g, then the new distribution with p.d.f.
$$h(x, y) = \lambda h_1(x, y) + (1 - \lambda)\, h_2(x, y), \quad 0 \leqslant \lambda \leqslant 1,$$

will also have the marginal p.d.f.'s f and g. Charlier and Wicksell (1923) first studied the moments of this kind of distribution and, in particular, the case when h_1 and h_2 correspond to $N(\mu, \nu, \sigma_1^2, \sigma_2^2, \rho)$ and $N(\mu', \nu', \sigma_1'^2, \sigma_2'^2, \rho')$, respectively. Hopkins and Clay (1963) used the latter model in a robustness study of Hotelling's T^2, and Bhattacharyya (1967) has examined the Pitman efficiency of some k-sample tests for alternatives of this kind.

9.5 A totally symmetrical class

Let $\phi(x)$ be the p.d.f.which differs from zero only when $X \geqslant 0$; then Cramér (1946, p. 219) has defined a totally symmetrical class of bivariate distributions with p.d.f.

$$h(x, y) = \frac{\phi\{(x^2 + y^2)^{1/2}\}}{2\pi(x^2 + y^2)^{1/2}}, \quad -\infty < x, y < \infty.$$

Let the random variable (X, Y) be transformed to polar co-ordinates by

$$X = R\cos\Theta, \quad Y = R\sin\Theta, \quad R \geqslant 0, \quad 0 < \Theta \leqslant 2\pi.$$

The p.d.f. of (R, Θ) is

$$\phi(r)/2\pi.$$

Hence R and Θ are independent random variables, the p.d.f. of R is $\phi(r)$, and the random variable Θ is uniformly distributed on a circle.

It can be seen that the independent bivariate normal (§ 10.6) and the bivariate Cauchy (§ 10.8) distributions are contained in this class.

Mardia (1967b) has studied the Pitman efficiency of a bivariate two-sample non-parametric test relative to Hotelling's T^2 for samples drawn from this type of population.

9.6 Concluding remarks

We have discussed various families of bivariate distributions. There are families which are satisfactory for problems arising in robustness studies, non-parametric methods, etc., but from the point of view of fitting, no system can yet be claimed to be adequate. Since the marginal distributions can be fitted successfully, we can always assume that the marginal distributions are specified, but there are many families with given margins, and so far no method has been proposed for deciding which family is the most appropriate. For each family, the margins specify the form of the bivariate distribution, but then the forms of the conditional distributions are not sufficiently elastic to allow for the wide variety of shapes of regression and higher array curves. Furthermore, for given margins no group of families has so far been studied which can give a broad choice of regression curves, scedastic curves, etc. What one perhaps needs is a family of uniform distributions with a variety of possible regression and scedastic curves, so that the data can be reduced to uniform margins, and after looking at the shapes of the observed regression and scedastic curves a member can then be selected. The last step is somewhat analogous to the method used for some univariate families for selecting a member on the basis of the observed values of β_1 and β_2. Such systems should at least be tested for their adequacy on the six observed distributions studied by Pretorius. The families developed after 1930 also need a thorough examination on the lines of Pretorius. Some of the families can certainly be excluded from further examination by comparing the behaviour of their conditional moment curves with those given by Pretorius for the observed distributions.

CHAPTER 10

SOME IMPORTANT BIVARIATE DISTRIBUTIONS

In this chapter we give some bivariate distributions whose marginal distributions are well-known univariate distributions. We have seen in earlier chapters that the marginal distributions do not determine the bivariate distributions uniquely. It seems natural, therefore, to restrict the class by imposing the condition that the conditional d.f's are of the same form as the margins, but there may not always be a distribution satisfying this requirement (see the bivariate Poisson distribution given in § 10.3). Generally, a meaningful bivariate distribution is produced by extending some special features of the univariate models. In some cases, the distributions arise naturally in theoretical sampling problems. A fairly complete list of bivariate distributions published before 1958 is given by Haight (1961, § 9). In Appendix 1, we list the bivariate distributions considered here and indicate the families to which they belong.

10.1 The bivariate binomial distribution

The probability function (p.f.) of the bivariate binomial distribution (the trinomial distribution) is given by

$$p(x, y) = \frac{n!}{x!\, y!\, (n-x-y)!}\, p_1^x\, p_2^y\, (1-p_1-p_2)^{n-x-y}, \tag{10.1.1}$$

where $0 < p_1, p_2 < 1$, $p_1 + p_2 < 1$, n is an integer, and $x, y = 0, 1, \ldots, n$ such that $x + y \leqslant n$. We have

$$\mu = np_1, \quad \sigma_1^2 = np_1(1-p_1), \quad \mu_{11} = -np_1 p_2,$$

and

$$\phi_{X,Y}(t_1, t_2) = (1 - p_1 - p_2 + p_1 e^{it_1} + p_2 e^{it_2})^n .$$

The conditional distributions are binomial, so that

$$\mathrm{E}(Y|x) = p_2(n-x)/(1-p_1), \quad \mathrm{var}(Y|x) = p_2(1-p_1-p_2)(n-x)/(1-p_1)^2.$$

10.2 The bivariate hypergeometric distribution

The p.f. of the bivariate hypergeometric distribution is given by

$$p(x, y) = \frac{\binom{Np_1}{x}\binom{Np_2}{y}\binom{N-Np_1-Np_2}{n-x-y}}{\binom{N}{n}} \tag{10.2.1}$$

where $x, y = 0, 1, \ldots, n$ such that $x \leqslant Np_1$, $y \leqslant Np_2$ and $n - x - y \leqslant N(1 - p_1 - p_2)$; N, n integers, $n \leqslant N$, $0 < p_1, p_2 < 1$ and $p_1 + p_2 \leqslant 1$. We have

$$\mu = np_1, \quad \sigma_1^2 = np_1(1-p_1)(N-n)/(N-1), \quad \mu_{11} = -np_1 p_2 (N-n)/(N-1).$$

The conditional distributions are hypergeometric, so that

$$\mathrm{E}(Y|x) = \frac{p_2(n-x)}{1-p_1}, \quad \mathrm{var}(Y|x) = \frac{N^2 p_2(1-p_1-p_2)}{N-Np_1-1}\left\{\frac{1}{4} - \left(\frac{N-x}{N-Np_1} - \frac{1}{2}\right)^2\right\}.$$

This distribution was introduced by Isserlis (1914) and he fitted the distribution to some data. K. Pearson (1924a, b) gives various properties. K. Pearson (1924b) has fitted this surface to an observed distribution of the number of cards of a certain suit in two hands at whist.

10.3 The bivariate Poisson distribution

The p.f. of the bivariate Poisson distribution is given by

$$p(x, y) = \exp(-\lambda_1 - \lambda_2 + \lambda_3) \cdot \frac{a^x b^y}{x!y!} \sum_{r=0}^{s} \frac{x^{(r)}}{a^r} \frac{y^{(r)}}{b^r} \frac{\lambda_3^r}{r!} \qquad (10.3.1)$$

where $s = \min(x, y)$, $a = \lambda_1 - \lambda_3$, $b = \lambda_2 - \lambda_3$, $\lambda_1 > \lambda_3 > 0$, $\lambda_2 > \lambda_3 > 0$, and $x^{(r)} = x(x-1)\ldots(x-r+1)$. The random variables X and Y have Poisson distributions with parameters λ_1 and λ_2, respectively, but the conditional distributions are not Poisson. In fact, no bivariate distribution exists having both marginal and conditional distributions of Poisson form (see Seshadri and Patil, 1964).

We have

$$\mu = \lambda_1, \quad \sigma_1^2 = \lambda_1, \quad \mu_{11} = \lambda_3, \quad 0 < \mu_{11} < \min(\lambda_1, \lambda_2),$$

and

$$\phi_{X,Y}(t_1, t_2) = \exp(-\lambda_1 - \lambda_2 - \lambda_3 + \lambda_1 e^{it_1} + \lambda_2 e^{it_2} + \lambda_3 e^{it_1 + it_2}).$$

Further,

$$\mathrm{E}(Y|x) = b + (\lambda_3/\lambda_1)x, \quad \mathrm{var}(Y|x) = b + \{a\lambda_3/\lambda_1^2\}x.$$

Campbell (1934) derived this distribution by considering the limiting distribution of a fourfold table, and the above explicit form is due to Aitken (1944). Holgate (1964) has obtained this distribution as follows. Let X_1, X_2 and X_3 be independent Poisson with parameters λ_1, λ_2 and λ_3 respectively. The joint distribution of $X = X_1 + X_3$ and $Y = X_2 + X_3$ is given by (10.3.1). A comprehensive review of the topic is given by Haight (1967, pp. 64–6).

10.4 The bivariate negative binomial distribution

The p.f. of the bivariate negative distribution is given by

$$p(x, y) = \frac{(x + y + k - 1)!}{x!\, y!\, k!} p_1^x p_2^y (1 - p_1 - p_2)^k, \qquad (10.4.1)$$

where $x, y = 0, 1, \ldots;\ k > 0,\ 0 < p_1, p_2 < 1$, and $p_1 + p_2 < 1$. We have

$$\mu = kp_1/a, \quad \sigma_1^2 = kp_1(1 - p_2)/a^2, \quad \mu_{11} = kp_1 p_2/a^2,$$

where $a = 1 - p_1 - p_2$, and

$$\phi_{X,Y}(t_1, t_2) = a^k/(1 - p_1 e^{it_1} - p_2 e^{it_2})^k.$$

The conditional distributions are again negative binomial, so that

$$\mathrm{E}(Y|x) = p_2(k + x)/(1 - p_2), \quad \mathrm{var}(Y|x) = p_2(k + x)/(1 - p_2)^2.$$

Guldberg (1934) introduced this distribution, and Lundberg (1940) first used it in connection with problems of accident proneness. Arbous and Kerrich (1951) form the distribution by mixing parameters of the bivariate Poisson distribution, and Bates and Neyman (1952a, b) have been led to this distribution by considering a model with gamma-type parameter mixing. Arbous and Kerrich (1951) and Bates and Neyman (1952a) have fitted this distribution to data related to accident-proneness. Neyman (1965) and Sibuya *et al.* (1964) have reviewed the subject. Edwards and Gurland (1961) have generalized (10.4.1) to

$$p(x, y) = A^{-k-x} B_1^{x-y} B_{12}^y \sum_{i=0}^{y} \frac{(-1)^{x+i}\, \Gamma(k + x + i)}{(x - y + i)!\,(y - i)!\, i!\, \Gamma(k)} \left[\frac{B_1 B_2}{AB_{12}}\right]^i$$

where $x \geqslant y$; $x, y = 0, 1, 2, \ldots;\ A > 0,\ k > 0$ and $B_1, B_2, B_{12} < 0$. For $B_{12} = 0$, it reduces to (10.4.1) and the authors compare the fits obtained from the two distributions. For $k = 1$, the distribution given by (10.4.1) becomes a bivariate geometric distribution.

10.5 The bivariate logarithmic distribution

The p.f. of the bivariate logarithmic distribution is given by

$$p(x,y) = \frac{(x+y-1)!}{x!\,y!} \frac{p_1^x p_2^y}{-\log(1-p_1-p_2)}, \qquad (10.5.1)$$

where $x, y = 0, 1, \ldots;\ x+y > 0,\ 0 < p_1, p_2 < 1$, and $p_1 + p_2 < 1$. We have

$$\mu = p_1/(ab), \quad \sigma_1^2 = (bp_1 - p_1 + ab)p_1/(ab)^2,$$
$$\mu_{11} = p_1 p_2 (b-1)/(ab)^2,$$

where $a = 1 - p_1 - p_2$ and $b = -\log a$.

$$\mathrm{E}(Y|x) = p_2/[(1-p_2)\{-\log(1-p_2)\}], \quad \text{for } x = 0,$$
$$= p_2 x/(1-p_2), \quad \text{for } x > 0.$$

The distribution was introduced by Khatri (1959). For further details, the reader is referred to Patil and Bildikar (1967) who also fit this distribution to data from the field of population and community ecology.

10.6 The bivariate normal distribution

For completeness we note that the probability density function (p.d.f.) of $N(0, 0, \sigma_1^2, \sigma_2^2, \rho)$ is given by

$$b(x,y;\rho) = \frac{1}{2\pi\sigma_1\sigma_2\sqrt{(1-\rho^2)}} \exp\left\{-\frac{1}{2(1-\rho^2)}\left(\frac{x^2}{\sigma_1^2} - \frac{2\rho xy}{\sigma_1\sigma_2} + \frac{y^2}{\sigma_2^2}\right)\right\}, \qquad (10.6.1)$$

where $-\infty < x, y < \infty$. The conditional distribution of Y given X is $N[\rho\sigma_2 x/\sigma_1, \sigma_2^2(1-\rho^2)]$, so that

$$\mathrm{E}(Y|x) = \rho\sigma_2 x/\sigma_1, \quad \mathrm{var}(Y|x) = \sigma_2^2(1-\rho^2).$$

Further,

$$\phi_{X,Y}(t_1, t_2) = \exp\{-\tfrac{1}{2}(\sigma_1^2 t_1^2 + 2\rho\sigma_1\sigma_2 t_1 t_2 + \sigma_2^2 t_2^2)\}.$$

10.7 Bivariate uniform distributions

Let S be a region in the Euclidean plane with area A; then the random variable (X, Y) is distributed uniformly over S if

$$h(x,y) = \frac{1}{A} \quad \text{for points } (x, y) \text{ in } S. \qquad (10.7.1)$$

For example, if the mass is uniformly distributed over an ellipse, then

$$h(x,y) = \frac{\sqrt{(ab - h^2)}}{\pi c}$$

for (x, y) in $ax^2 + by^2 + 2hxy < c$, $a > b > 0$, $h^2 < ab$ and $c > 0$.

We have already studied Plackett's uniform distribution given by (8.1.6) and Morgenstern's uniform distribution given by (9.2.3), both of which have uniform margins.

We now derive the uniform distribution on a sphere. Let (x, y, z) be transformed to polar co-ordinates by

$$x = r\cos\theta\sin\phi, \quad y = r\sin\theta\sin\phi, \quad z = r\cos\phi. \quad (10.7.2)$$

On calculating the Jacobian of the transformation, we get

$$dx\,dy\,dz = r^2 dr \times \sin\phi\,d\theta\,d\phi.$$

Hence, by analogy with the uniform distribution on a circle, we can say that the mass on the unit sphere is uniformly distributed if the p.d.f. of Θ and Φ is

$$h(\theta,\phi) = k\sin\phi\,d\theta\,d\phi, \quad 0 < \phi \leqslant 2\pi, \quad 0 < \theta < \pi. \quad (10.7.3)$$

It can easily be verified that $k = 1/(4\pi)$.

10.8 The bivariate Cauchy distribution

The p.d.f. of the bivariate Cauchy distribution is

$$h(x, y) = (2\pi)^{-1}\, c[c^2 + x^2 + y^2]^{-3/2}, \quad (10.8.1)$$

where $-\infty < x, y < \infty$ and $c > 0$.

The distributions of X and Y are both Cauchy. The conditional p.d.f. of Y given X is

$$\tfrac{1}{2}(c^2 + x^2)/(c^2 + x^2 + y^2)^{3/2},$$

so that $Y/\sqrt{\{\tfrac{1}{2}(x^2 + c^2)\}}$ has a Student distribution with two degrees of freedom. Consequently, $\mathrm{E}(Y|x)$ does not exist. Thus, for this distribution μ, ν, σ_1, σ_2 and μ'_{11} do not exist. The curves of equal probability for this distribution are circles.

This distribution can be derived by considering a physical model. A source O emanates radioactive particles which hit a two-dimensional screen. Suppose the screen is hit at the point P, and let M be the foot of the perpendicular from O to the screen. Consider the axes OX and OY in a plane parallel to the screen. If $r = OP$, $\angle POM = \phi$, and θ is the angle made by the projection of OP with X-axis, then the Cartesian co-ordinates (x, y, z) of P are given by the polar transformation (10.7.2). Hence

$$x = c\tan\phi\cos\theta, \quad y = c\tan\phi\sin\theta,$$

where $c = OM$. On assuming that all directions of emission are equally likely, the p.d.f. of (θ, ϕ) is given by (10.7.3), and so (X, Y) has the p.d.f. given by (10.8.1).

The distribution is a particular case of the bivariate Student's t first given by K. Pearson (1923b). This form is also given by Morgenstern (1956) and the above motivation is due to Rao (1965, pp. 137–8).

10.9 The bivariate beta distribution

The p.d.f. of the bivariate beta distribution (the Filon–Isserlis surface) is given by

$$h(x, y) = \frac{\Gamma(p_1 + p_2 + p_3)}{\Gamma(p_1)\,\Gamma(p_2)\,\Gamma(p_3)}\, x^{p_1-1}\, y^{p_2-1}(1 - x - y)^{p_3-1}; \qquad (10.9.1)$$

where $x \geqslant 0, y \geqslant 0$, $x + y \leqslant 1$ and $p_1, p_2, p_3 > 0$. We have

$$\mu = p_1/a, \quad \sigma_1^2 = p_1(p_2 + p_3)/\{a^2(a + 1)\},$$

and

$$\rho = -\sqrt{[\{p_1 p_2\}/\{(p_1 + p_3)(p_2 + p_3)\}]},$$

where $a = p_1 + p_2 + p_3$. The random variable X has the beta distribution with index parameters p_1 and $p_2 + p_3$. It can be seen that the conditional distribution of $Y/(1 - x)$ is beta with parameters p_2 and p_3. Hence

$$\mathrm{E}(Y|x) = p_2(1 - x)/(p_2 + p_3),$$
$$\mathrm{var}(Y|x) = p_2 p_3 (1 - x)^2/\{(p_2 + p_3)^2 (p_2 + p_3 + 1)\}.$$

It should be noted that this distribution has the following desirable property. Suppose X_1, X_2 and X_3 are independent r.v's having gamma distributions with index parameters p_1, p_2 and p_3; then Johnson (1960) has shown that

$$X = X_1/(X_1 + X_2 + X_3), \quad Y = X_2/(X_1 + X_2 + X_3)$$

has the distribution given by (10.9.1).

This distribution is due to L.N.G. Filon and L. Isserlis (K. Pearson 1923a). K. Pearson (1924b) fitted the distribution to data of whist correlation. Johnson (1960) has suggested this distribution as an approximation to the multinomial distribution. Wilks (1962, pp. 177–82) has given some properties of this distribution.

The bivariate Fisher–Hsu–Roy distribution of the characteristic roots of the sample covariance matrix (Kendall, 1957, p. 92) has the p.d.f.

$$\text{const.}\ (\lambda_1 \lambda_2)^a \{(1-\lambda_1)(1-\lambda_2)\}^b (\lambda_1 - \lambda_2), \quad 0 < \lambda_2 < \lambda_1 < 1.$$

On using the transformation $X = \sqrt{(\lambda_1 \lambda_2)}$ and $Y = \sqrt{\{(1-\lambda_1)(1-\lambda_2)\}}$, we find that the resulting distribution is of the form (10.9.1).

Olkin and Rubin (1964) have studied various additional multivariate beta distributions.

10.10 Bivariate gamma distributions

We shall consider here three bivariate gamma distributions. First with the p.d.f.

$$h_1(x, y) = \frac{C^{-\frac{1}{2}(p-1)}}{\Gamma(p)(1-C)} (xy)^{\frac{1}{2}(p-1)} \exp\left(-\frac{x+y}{1-C}\right) I_{p-1}\left[\frac{2(Cxy)^{1/2}}{1-C}\right], \tag{10.10.1}$$

where $x \geqslant 0$, $y \geqslant 0$, $p > 0$, $0 \leqslant C < 1$ and

$$I_k(z) = \sum_{r=0}^{\infty} \frac{(\frac{1}{2}z)^{k+2r}}{r!\,\Gamma(k+r+1)}$$

which is a modified Bessel function of the first kind and the kth order.

The random variables X and Y both have gamma distributions with the same index parameter p. We have $\mu = p$, $\sigma_1^2 = p$, $\rho = C$,

$$\mathrm{E}(Y|x) = Cx + p(1-C), \quad \mathrm{var}(Y|x) = (1-C)\{2Cx + p(1-C)\}$$

and

$$\phi_{X,Y}(t_1, t_2) = [(1-t_1)(1-t_2) - Ct_1 t_2]^{-p}.$$

Let (X_i, Y_i), $i = 1, \ldots, n$, be a random sample from a bivariate normal population with zero means. By considering $X = \frac{1}{n}\Sigma X_i^2$ and $Y = \frac{1}{n}\Sigma Y_i^2$, Wicksell (1933) derived the characteristic function of this distribution. The explicit form (10.10.1) is due to Kibble (1941). Rice (1944, 1945) has given an application to noise theory, and Vere-Jones (1967) has shown it to be infinitely divisible. For $p = 1$ in (10.10.1), we get a bivariate exponential distribution. Moran (1967) has studied certain tests of independence for samples drawn from this bivariate exponential population.

Cherian (1941) constructed a bivariate gamma distribution as follows. Let X_1, X_2 and X_3 have gamma distributions with index parameters p_1, p_2 and p_3 respectively. Then $X = X_1 + X_3$ and $Y = X_2 + X_3$ have a bivariate gamma distribution. In fact,

$$h_2(x, y) = \frac{e^{-(x+y)}}{\prod_{i=1}^{3} \Gamma(p_i)} \int_0^{\min(x,y)} z^{p_3-1}(x-z)^{p_1-1}(y-z)^{p_2-1} e^z \, dz, \tag{10.10.2}$$

where $x, y > 0$ and $p_1, p_2, p_3 > 0$. The p.d.f's of X and Y are gamma distributions with index parameters $p_1 + p_3$ and $p_2 + p_3$, respectively. We have

$$\rho = p_3/\sqrt{\{(p_1 + p_3)(p_2 + p_3)\}}, \quad \mathrm{E}(Y|x) = p_2 + \{p_3/(p_1 + p_3)\}x.$$

David and Fix (1961) have studied the rank correlation and regression for samples from this distribution. An account of bivariate gamma distributions is given by Moran (1967).

Another bivariate gamma distribution of interest has p.d.f.

$$h_3(x, y) = \frac{a^{p+q}}{\Gamma(p)\,\Gamma(q)} x^{p-1}(y-x)^{q-1} e^{-ay}, \tag{10.10.3}$$

where $y > x > 0$ and $a, p, q > 0$. The p.d.f. of X is

$$\psi(x; p, a) = \frac{a^p}{\Gamma(p)} x^{p-1} e^{-ax},$$

and the p.d.f. of Y is $\psi(y; p+q, a)$. We have $\mu = p/a$, $\sigma_1^2 = p/a^2$, $\rho = \sqrt{\{p/(p+q)\}}$. The conditional p.d.f. of $U = Y - x$, given x, is $\psi(u; q, a)$, and the conditional distribution of $V = X/y$, given y, is beta with parameters p and q. Hence

$$\mathrm{E}(Y|x) = (q/a) + x, \quad \mathrm{var}(Y|x) = q/a^2,$$

$$\mathrm{E}(X|y) = \{p/(p+q)\}y, \quad \mathrm{var}(X|y) = \{pq/(p+q)^2(p+q+1)\}y^2.$$

McKay (1934) derived this distribution as follows. Let $(X_1, \ldots, X_N)$ be a random sample from a normal population. Suppose S_N^2 is the sample variance, and let S_n^2 be the variance in a sub-sample of size n from $(X_1, \ldots, X_N)$. The joint distribution of S_N^2 and S_n^2 is of the form (10.10.3). Mihram and Hultquist (1967) describe X as a warning-time variable and Y as the failure-time of the component on test. They also generalized the distribution to

$$|c| x^{p-1}(y-x)^{q-1} y^{bc-p-q} e^{-(y/a)^c} / a^{bc}\,\Gamma(b)\,\beta(p, q),$$

where $0 < x < y < \infty$, $a, b, p, q > 0$.

10.11 Bivariate exponential distributions

On taking the index parameters to be unity in the first two bivariate gamma distributions given above, we obtain the corresponding two bivariate exponential distributions. In addition to these, other exponential distributions are of interest.

Marshall and Olkin (1967) have introduced a bivariate exponential distribution with d.f.

$$H(x, y) = 1 - \exp\{-(\lambda_1 + \lambda_3)x\} - \exp\{-(\lambda_2 + \lambda_3)y\} + \exp\{-\lambda_1 x - \lambda_2 y + \lambda_3 \max(x, y)\}, \qquad (10.11.1)$$

where $x, y > 0$ and $\lambda_1, \lambda_2, \lambda_3 > 0$. We have

$$H(x,\infty) = 1 - \exp\{-(\lambda_1 + \lambda_3)x\}, \quad H(\infty, y) = 1 - \exp\{-(\lambda_2 + \lambda_3)y\}$$

which are univariate exponential distributions. We have

$$\mu = 1/(\lambda_1 + \lambda_3), \quad \sigma_1^2 = 1/(\lambda_1 + \lambda_3)^2, \quad \rho = \lambda_3/\lambda,$$

and $\phi_{X,Y}(t_1, t_2)$

$$= \frac{\lambda_1 + \lambda_3}{\lambda_1 + \lambda_3 - it_1} + \frac{\lambda_2 + \lambda_3}{\lambda_2 + \lambda_3 - it_2} - \frac{t_1 t_2 (\lambda + \lambda_3 - it_1 - it_2)}{(\lambda - it_1 - it_2)(\lambda_1 + \lambda_3 - it_1)(\lambda_2 + \lambda_3 - it_2)} - 1,$$

where $\lambda = \lambda_1 + \lambda_2 + \lambda_3$. The authors have given various meaningful derivations. By considering models in which a two-component system fails to function after occurrence of a shock to each or both components, they obtain (10.11.1). Further, on assuming the lack of memory property

$$P(X > s_1 + t, Y > s_2 + t \mid X > t, Y > t) = P(X > s_1, Y > s_2)$$

and exponential margins, (10.11.1) is again obtained.

On transforming the random variable (X, Y) to (X', Y') by $X' = X^{1/\beta}$, $Y = Y^{1/\nu}$, we get a bivariate Weibull distribution.

Gumbel (1960) has studied the bivariate distribution with p.d.f.

$$h(x, y) = \{(1 + ax)(1 + ay) - a\}\, e^{-x-y-axy}, \quad x > 0, y > 0, a > 0, \qquad (10.11.2)$$

which has exponential margins, but no meaningful derivation is known. We have

$$E(Y|x) = (1 + a + ax)/(1 + ax)^2,$$

$$\operatorname{var}(Y|x) = \{(1 + 2a - a^2) + 2a(1 + a)x + a^2x^2\}/(1 + ax)^4,$$

and
$$\rho = -1 + \int_0^\infty \frac{e^{-y}}{1 + ay}\, dy.$$

10.12 The bivariate Pareto distribution

The p.d.f. of the bivariate Pareto distribution is given by

$$h(x, y) = p(p + 1)(ab)^{p+1}/(bx + ay - ab)^{p+2}, \qquad (10.12.1)$$

where $x > a > 0$, $y > b > 0$, $p > 0$.

The marginal p.d.f. of X is

$$f(x; a, p) = pa^p/x^{p+1}, \quad x > a > 0,$$

which is the univariate Pareto distribution. The conditional p.d.f. of Y given X is

$$a(p + 1)(bx)^{p+1}/(bx + ay - ab)^{p+2},$$

which is again of Pareto form but with displaced origin. We have

$$\mu = ap/(p - 1), \quad p > 1; \quad \sigma_1^2 = a^2p/\{(p - 1)^2 (p - 2)\}, \; p > 2;$$

$$\rho = 1/p, \quad p > 2;$$

$$\mathrm{E}(Y|x) = b + \{(bx)/(ap)\}, \quad \mathrm{var}(Y|x) = b^2x^2(p + 1)/\{a^2(p - 1)p^2\}.$$

The curves of equal probability are straight lines. The univariate Pareto distribution has the property that if X has the p.d.f. $f(x; a, p)$ then the r.v. X truncated on the left at $X = a'$, $a' > a$ has the p.d.f. $f(x; a', p)$. The random variable (X', Y') given by $X' = b(X - ac)$, $Y' = a(Y - b + bc)$, with $0 < c < 1$, has the p.d.f. of the form

$$h(x', y'; a, b, p) = p(p + 1)(a + b)^p/(x' + y')^{p+2},$$

where $x' > a > 0$, $y' > b > 0$ and $p > 0$. The p.d.f. of the r.v. (X', Y'), truncated on the left at $X' = a'$, $Y' = b'$ with $a' > a$ and $b' > b$, is $h(x', y'; a', b', p)$. Further, we have

$$\log \mathrm{P}(X > x) = \frac{1}{p} \log (x/a),$$

and this property is again preserved by the bivariate Pareto distribution as

$$\log \mathrm{P}(X > x, Y > y) = \frac{1}{p} \log (a^{-1}x + b^{-1}y - 1).$$

For further details, the reader is referred to Mardia (1962).

10.13 The bivariate Student distribution

The p.d.f. of the bivariate Student's t-distribution is given by

$$h(x,y)$$
$$= \{2\pi\sigma_1\sigma_2\sqrt{(1-\rho^2)}\}^{-1}\frac{f}{f-2}\left[1+\frac{1}{(f-2)(1-\rho^2)}\left\{\frac{x^2}{\sigma_1^2}-\frac{2\rho xy}{\sigma_1\sigma_2}+\frac{y^2}{\sigma_2^2}\right\}\right]^{-\frac{1}{2}(f+2)},$$

(10.13.1)

where $-\infty < x,\ y < \infty$, $f > 0$. The r.v.'s X/σ_1 and Y/σ_2 both have the Student's t-distribution with f degrees of freedom (d. of fr.). We have

$$E(X) = 0, \quad \mathrm{var}(X) = \sigma_1^2, \quad \mathrm{corr}(X, Y) = \rho.$$

The random variable

$$U = \frac{\sigma_1\sqrt{(f+1)}}{\sqrt{\{(f-2)\sigma_1^2+x^2\}}}\left(Y-\rho\frac{\sigma_2}{\sigma_1}x\right)\Big/\{\sigma_2\sqrt{(1-\rho^2)}\}$$

for given x has a t-distribution with $f+1$ d. of fr. Hence,

$$E(Y|x) = \rho\sigma_2 x/\sigma_1, \quad \mathrm{var}(Y|x) = \sigma_2^2(1-\rho^2)\{(f-2)\sigma_1^2+x^2\}/\{(f-1)\sigma_1^2\}.$$

It can be seen that the curves of equal probability are ellipses. For $f = 1$ and $\rho = 0$, the distribution reduces to the bivariate Cauchy distribution (§ 10.8). As $f \to \infty$, the distribution tends to the bivariate normal.

This distribution was introduced by K. Pearson (1923b) for graduation purposes. K. Pearson (1924b) fitted the distribution to data of whist correlation. Dunnett and Sobel (1954) found that it arises naturally in multiple decision problems concerned with the ranking of normal populations with a common unknown variance. Gupta (1963a, b) has reviewed the topic.

Siddiqui (1967) has studied the bivariate t-distribution defined as the distribution of (t_1, t_2), t_1 and t_2 being t-statistics for a random sample from $N(0, 0, 1, 1, \rho)$.

10.14 The bivariate F-distribution

The p.d.f. of the bivariate F-distribution is given by

$$h(x,y) = \Gamma(\tfrac{1}{2}\nu)\,\nu_0^{-\frac{1}{2}\nu}\prod_{i=0}^{2}\{\nu_i^{\frac{1}{2}\nu_i}/\Gamma(\tfrac{1}{2}\nu_i)\}\frac{x^{\frac{1}{2}\nu_1-1}\,y^{\frac{1}{2}\nu_2-1}}{\left(1+\dfrac{\nu_1 x+\nu_2 y}{\nu_0}\right)^{\frac{1}{2}\nu}},$$

$$0 < x,\ y < \infty,$$

(10.14.1)

where $\nu = \nu_0 + \nu_1 + \nu_2$. The random variable X has an F-distribution with degrees of freedom (d. of fr.) ν_1 and ν_0, and Y has an F-distribution with d. of fr. ν_2 and ν_0. We have

$$\mu = \nu_0/(\nu_0 - 2), \quad \nu_0 > 2,$$
$$\sigma_1^2 = 2\nu_0^2(\nu_0 + \nu_1 - 2)/\{\nu_1(\nu_0 - 2)^2(\nu_0 - 4)\}, \quad \nu_0 > 4,$$

and
$$\rho = \sqrt{[\nu_1\nu_2/\{(\nu_0 + \nu_1 - 2)(\nu_0 + \nu_2 - 2)\}]}, \quad \nu_0 > 4.$$

Further, the conditional p.d.f. of $(\nu_0 + \nu_1)Y/(\nu_0 + \nu_1 x)$ for given x has an F-distribution with d. of fr. ν_2 and $\nu_0 + \nu_1$, so that

$$\mathrm{E}(Y|x) = (\nu_0 + \nu_1 x)/(\nu_0 + \nu_1 - 2)$$
$$\mathrm{var}\,(Y|x) = [2(\nu - 2)/\{\nu_2(\nu_0 + \nu_1 - 2)^2(\nu_0 + \nu_1 - 4)\}](\nu_0 + \nu_1 x)^2.$$

The curves of equal probability are straight lines. This distribution arises naturally in the simultaneous analysis of variance. Let X_0, X_1 and X_2 be independent r.v's having chi-squared distributions with d. of fr. ν_0, ν_1 and ν_2 respectively; then the random variable (X, Y) given by

$$X = \frac{X_1/\nu_1}{X_0/\nu_0}, \quad Y = \frac{X_2/\nu_2}{X_0/\nu_0}$$

has the distribution given by (10.14.1), as pointed out by Ghosh (1955). In the randomized block design, X_0 is the error sum of squares and X_1, X_2 are the sums of squares for the other two components.

If we put $\nu_1 = \nu_2 = 1$, and transform the random variable (X, Y) to (U, V) by $U = X^2$ and $V = Y^2$, then (10.14.1) reduces to a particular case ($\rho = 0$) of the bivariate Student's t-distribution given by (10.13.1). The r.v. (U, V) defined by

$$U = \nu_1 X/(\nu_0 + \nu_1 X), \quad V = \nu_2 Y/(\nu_0 + \nu_2 Y)$$

does not have the bivariate beta distribution given by (10.9.1), although the marginal distributions of U and V are of beta type.

10.15 The bivariate Type II distribution

The p.d.f. of the bivariate Type II distribution is

$$h(x, y) = \{2\pi\sigma_1\sigma_2\sqrt{(1 - \rho^2)}\}^{-1}\frac{n+1}{n+2}\left[1 - \frac{1}{2(n+2)(1-\rho^2)}\right.$$
$$\left.\times\left\{\frac{x^2}{\sigma_1^2} - \frac{2\rho xy}{\sigma_1\sigma_2} + \frac{y^2}{\sigma_2^2}\right\}\right]^n, \tag{10.15.1}$$

where $-c\sigma_1 < x < c\sigma_1$, $-c\sigma_2 < y < c\sigma_2$, $c^2 = 2(n+2)$ and $n > 0$. The marginal p.d.f. of X is

$$\frac{(n+1)\,\beta(n+1,\tfrac{1}{2})}{\pi\sigma_1\, c}\left(1 - \frac{x^2}{c^2\sigma_1^2}\right)^{n+1}, \quad -c\sigma_1 < x < c\sigma_1,$$

which is a Type II distribution. The p.d.f. of Y is of a similar form. We have

$$\mathrm{E}(Y|x) = \rho\,\frac{\sigma_2}{\sigma_1}\,x, \quad \mathrm{var}(Y|x) = \sigma_2^2(1-\rho^2)\left\{1 + \frac{1}{2n+3}\left(1 - \frac{x^2}{\sigma_1^2}\right)\right\}.$$

It can be seen that the curves of equal probability are ellipses. The distribution was introduced and studied by K. Pearson (1923b). Sagrista (1952) has also given certain properties.

10.16 Rhodes' distribution

The p.d.f. of Rhodes' distribution is

$$h(x,y) = \frac{p^{p+1}p'^{p'+1}\,(ab'-a'b)}{e^{p+p'}\Gamma(p+1)\Gamma(p'+1)\,aa'bb'}\,e^{-lx-my}\left(1-\frac{x}{a}+\frac{y}{b}\right)^p\left(1+\frac{x}{a'}-\frac{y}{b'}\right)^{p'} \tag{10.16.1}$$

where the region of non-zero density is defined by

$$S = \left\{(x,y):\ 1 - \frac{x}{a} + \frac{y}{b} > 0,\ 1 + \frac{x}{a'} - \frac{y}{b'} > 0\right\}$$

and a, b, a', b', p and p' are positive, and $ab' > a'b$. The region S can also be written as

$$0 < \frac{b}{a}x - b < y < \frac{b'}{a'}x + b' < \infty.$$

The condition $ab' > a'b$ ensures that the two lines forming the boundaries intersect in the third quadrant, and the region consists of the part of the plane enclosed by the lines containing the origin. The situation is reversed for the case $ab' < a'b$. Let

$$A = (ab' - a'b)/aa'bb'.$$

The marginal p.d.f. of Y is given by

$$g(y) = \text{const.}\ e^{-pa'u}\,u^{p+p'+1}\int_0^1 t^{p'}(1-t)^p\,e^{-laa'ut}\,dt,$$

where
$$u = \frac{1}{a} + \frac{1}{a'} + Ay.$$

Or, equivalently

$$g(y) = \text{const. } e^{-pa'u} B(R, s'),$$

where $R = p + p' + 2$, $s' = p' + 1$ and

$$B(R, s') = u^{R-1} - \frac{aa'ls'}{R} u^R + \frac{(aa'l)^2}{2!} \frac{s'(s'+1)}{R(R+1)} u^{R+1} - \ldots .$$

Further, we have

$$\text{E}(Y|x) = a'\left(\frac{y}{b'} - 1\right) + \frac{as'}{R} \frac{B(R+1, s'+1)}{B(R, s')} .$$

The distribution has a mode which is given by

$$1 - \frac{x}{a} + \frac{y}{b} = -Ap\bigg/\left(\frac{m}{a'} + \frac{l}{b'}\right),$$

$$1 + \frac{x}{a'} - \frac{y}{b'} = -Ap'\bigg/\left(\frac{m}{a} + \frac{l}{b}\right).$$

The distance of the mean from the mode is

$$\mu = \frac{1}{A}\left(\frac{1}{pb'} + \frac{1}{p'b}\right), \quad \nu = \frac{1}{A}\left(\frac{1}{pa'} + \frac{1}{p'a}\right).$$

Let $s = p + 1$. We have

$$\sigma_1^2 = \frac{1}{A^2}\left(\frac{s}{p^2b'^2} + \frac{s'}{p'^2b^2}\right), \quad \sigma_2^2 = \frac{1}{A^2}\left(\frac{s}{p^2a'^2} + \frac{s'}{p'^2a'^2}\right)$$

and

$$\mu_{11} = \frac{1}{A^2}\left(\frac{s}{p^2a'b'} + \frac{s'}{p'^2ab}\right).$$

The characteristics of this surface are as follows: (i) the marginal distributions tend to gamma distributions as the correlation approaches unity; (ii) the regression curves tend to linearity as the correlation approaches unity; and (iii) var $(Y|x)$ increases as x increases.

Rhodes (1923) has given various other properties and fitted this distribution to an observed distribution of barometric heights recorded simultaneously at Southampton and Laudale, Argyll. The form (10.16.1) was suggested by the fact that the original data are bounded, roughly, by two intersecting straight lines, the exponential term being introduced to ensure that, for x, y large, $h(x, y)$ should tend to zero. Hill (1954) has studied the distribution of the regression coefficients in samples from this population.

APPENDIX

List of bivariate distributions with their relation to families

Section	Name	Probability function	Family
10.1	Binomial	$\frac{n!}{x!\,y!\,(n-x-y)!} p_1^x p_2^y (1-p_1-p_2)^{n-x-y}$ $x, y = 0, 1, \ldots, n,\ x+y \leqslant n.$	Darmois–Koopman–Pitman [§9.3]
10.2	Hypergeometric	$\binom{Np_1}{x}\binom{Np_2}{y}\binom{N-Np_1-Np_2}{n-x-y}\Big/\binom{N}{n}$ $x, y = 0, 1, \ldots, n, \quad x \leqslant Np_1, \quad y \leqslant Np_2,$ $n-x-y \leqslant N(1-p_1-p_2).$	
10.3	Poisson	$\exp(-\lambda_1-\lambda_2+\lambda_3)\frac{a^x b^y}{x!\,y!}\sum_{r=0}^{s}\frac{x^{(r)}}{a^r}\frac{y^{(r)}}{b^r}\frac{\lambda_3^r}{r!}$ $x, y = 0, 1, \ldots, \infty, \quad s = \min(x, y),$ $a = \lambda_1-\lambda_3, \quad b = \lambda_2-\lambda_3$	Arnold [§9.1]
10.4	Negative binomial	$\frac{(x+y+k-1)!}{x!y!k!} p_1^x p_2^y (1-p_1-p_2)^k$ $x, y = 0, 1, \ldots, \infty.$	Darmois–Koopman–Pitman [§9.3]
10.5	Logarithmic	$\frac{(x+y-1)!}{x!\ y!}\ \frac{p_1^x\ p_2^y}{-\log(1-p_1-p_2)}$ $x, y = 0, 1, \ldots, \infty, \quad x+y > 0$	Darmois–Koopman–Pitman [§9.3]

List of bivariate distributions with their relation to families (*continued*)

Section	Name	Probability density function	Family
10.6	Normal	$\{2\pi\sqrt{(1-\rho^2)}\}^{-1}\exp\left\{-\frac{1}{2(1-\rho^2)}(x^2-2\rho xy+y^2)\right\}$ $-\infty < x,\ y < \infty.$	K. Pearson [§2.1], van Uven [§2.2], Edgeworth [§3.1], Johnson [§4.2], Steffensen [§5.1], Narumi [§6.1], Fisher–Barrett–Lampard–Lancaster [§7.3], Arnold [§9.1], Darmois–Koopman–Pitman [§9.3]
10.7	Uniform (a) Morgenstern (b) Plackett	$1+\alpha(2x-1)(2y-1),\quad 0 \leqslant x, y \leqslant 1.$ $\frac{\psi\{(\psi-1)(x+y-2xy)+1\}}{[\{1+(\psi-1)(x+y)\}^2-4\psi(\psi-1)xy]^{3/2}},\quad 0 \leqslant x,\ y \leqslant 1.$	K. Pearson [§2.1], van Uven [§2.2], Morgenstern [§9.2] Plackett [§8.1]
10.8	Cauchy	$(2\pi)^{-1}c[c^2+x^2+y^2]^{-3/2},\quad -\infty < x,\ y < \infty.$	K. Pearson [§2.1], van Uven [§2.2], Narumi [§6.1], Cramér [§9.5]
10.9	Beta	$\frac{\Gamma(p_1+p_2+p_3)}{\Gamma(p_1)\Gamma(p_2)\Gamma(p_3)}x^{p_1-1}y^{p_2-1}(1-x-y)^{p_3-1}$ $x, y \geqslant 0,\quad x+y \leqslant 1.$	van Uven [§2.2], Narumi [§6.1]

List of bivariate distributions with their relation to families (continued)

Section	Name	Probability density function	Family
10.10	Gamma (a) Wicksell–Kibble	$\frac{1}{C^{(p-1)}\Gamma(p)} e^{-\frac{x+y}{1-C}} \sum_{k=0}^{\infty} \frac{(Cxy)^{p+k-1}}{k!\,\Gamma(p+k)\,(1-C)^{p+2k}}$ $x, y \geqslant 0.$	Fisher–Barrett–Lampard–Lancaster [§7.3]
	(b) Cherian	$\frac{e^{-(x-y)}}{\Gamma(p_1)\Gamma(p_2)\Gamma(p_3)} \int_0^{\min(x,y)} z^{p_3-1}(x-z)^{p_1-1}(y-z)^{p_2-1} e^{z}\, dz,$ $x,y \geqslant 0.$	Arnold [§9.1]
	(c) McKay	$\frac{a^{p+q}}{\Gamma(p)\,\Gamma(q)} x^{p-1}(y-x)^{q-1} e^{-ay}, \quad y > x > 0.$	van Uven [§2.2]
10.11	Exponential (a) Marshall & Olkin	$P(X > x,\ Y > y) = \exp\{-\lambda_1 x - \lambda_2 y + \lambda_3 \max(x, y)\},$ $x,y > 0.$	Arnold [§9.1]
	(b) Gumbel	$\{(1+ax)(1+ay)-a\} \exp(-x-y-axy), \quad x,y > 0.$	K. Pearson [§2.1]
10.12	Pareto	$p(p+1)(ab)^{p+1}/(bx+ay-ab)^{p+2}, \quad x > a,\ y > b.$	K. Pearson [§2.1], van Uven [§2.2], Narumi [§6.1]
10.13	Student's t	$\{2\pi\sqrt{(1-\rho^2)}\}^{-1}\{f/(f-2)\}$ $\times \left[1 + \frac{1}{(f-2)(1-\rho^2)}(x^2 - 2\rho xy + y^2)\right]^{-(f+2)/2}$ $-\infty < x,\ y < \infty.$	K. Pearson [§2.1], van Uven [§2.2], Narumi [§6.1]

List of bivariate distribution with their relation to families (continued)

Section	Name	Probability density function	Family
10.14	F	$\Gamma(\frac{1}{2}\nu)\,\nu_0^{-\frac{1}{2}\nu}\prod_{i=0}^{2}\{\nu^{\frac{1}{2}\nu_i}/\Gamma(\frac{1}{2}\nu_i)\}\ \dfrac{x^{\frac{1}{2}\nu_1-1}\,y^{\frac{1}{2}\nu_2-1}}{\left(1+\dfrac{\nu_1 x+\nu_2 y}{\nu_0}\right)^{\frac{1}{2}\nu}}$ $x, y > 0, \quad \nu = \nu_0 + \nu_1 + \nu_2.$	van Uven [§ 2.2], Narumi [§ 6.1]
10.15	Type II	$\{2\pi\sqrt{(1-\rho^2)}\}^{-1}\{(n+1)/(n+2)\}$ $\times\left[1-\dfrac{1}{2(n+2)(1-\rho^2)}(x^2-2\rho xy+y^2)\right]^n,$ $\lvert x\rvert, \lvert y\rvert < \sqrt{\{2(n+2)\}}.$	K. Pearson [§ 2.1], van Uven [§ 2.2], Narumi [§ 6.1]
10.16	Rhodes'	$\dfrac{p^{p+1}p'^{p'+1}(ab'-a'b)}{e^{p+p'}\Gamma(p+1)\,\Gamma(p'+1)\,aa'\,bb'}\,e^{-lx-my}$ $\times\left(1-\dfrac{x}{a}+\dfrac{y}{b}\right)^{p}\left(1+\dfrac{x}{a'}-\dfrac{y}{b'}\right)^{p'},\ 1-\dfrac{x}{a}+\dfrac{y}{b}>0, 1+\dfrac{x}{a'}-\dfrac{y}{b'}>0.$	K. Pearson [§ 2.1], Steffensen [§ 5.1]

REFERENCES AND AUTHOR INDEX

Sections in which authors are cited are given in brackets following each entry.

Aitken, A.C. (1944). *Statistical Mathematics*. 3rd edn. Edinburgh and London: Oliver and Boyd. [10.3]

Arbous, A.G. and Kerrich, J.E. (1951). Accident statistics and the concept of accident-proneness. *Biometrics*, **7**, 340–432. [10.4]

Arnold, B.C. (1967). A note on multivariate distributions with specified marginals. *J. Amer. Statist. Ass.*, **62**, 1460–1. [9.1]

Barrett, J.F. and Lampard, D.G. (1955). An expansion for some second-order probability distributions and its application to noise problems. *IRE Trans. Prof. Group on Information Theory, IT*, **1**, 10–15. [7.3, 7.4]

Barton, D.E. and Dennis, K.E. (1952). The conditions under which Gram–Charlier and Edgeworth curves are positive definite and unimodal. *Biometrika*, **39**, 425–7. [3.5]

Bates, G.E. and Neyman, J. (1952a,b). Contributions to the theory of accident proneness, I, II. *Univ. Calif. Publications in Statistics*, **1**, 215–54, 255–75. [10.4]

Berkson, J. (1951). Why I prefer logits to probits. *Biometrics*, **7**, 327–39. [4.3c]

Bhattacharyya, G.K. (1967). Asymptotic efficiency of multivariate normal score test. *Ann. Math. Statist.*, **38**, 1753–8. [9.4]

Bhuchongkul, S. (1964). A class of nonparametric tests for independence in bivariate populations. *Ann. Math. Statist.*, **35**, 138–49. [9.1]

Bildikar, S. and Patil, G.P. (1968). Multivariate exponential-type distributions. *Ann. Math. Statist.*, **39**, 1316–26. [9.3]

Box, G.E.P. and Muller, M.E. (1958). A note on the generation of normal deviates. *Ann. Math. Statist.*, **29**, 610–11. [4.3d]

Campbell, J.T. (1934). The Poisson correlation function. *Proc. Edin. Math. Soc.*, Series **2**, **4**, 18–26. [10.3]

Chambers, J.M. (1967). On methods of asymptotic approximation for multivariate distributions. *Biometrika*, **54**, 367–83. [3.1]

Charlier, C.V.L. (1914). Contributions to the mathematical theory of statistics, **6**. The correlation function of Type A. *Ark. f. Math., Astr. o. Fysik*, **9**, No. 26, 1–18. [3.1]

Charlier, C.V.L. and Wicksell, S.D. (1923). On the dissection of frequency functions. *Ark. f. Math., Astr. o. Fysik*, **18**, No. 6, 1–64. [9.4]

Chatterjee, S.K. (1960). Sequential tests for the bivariate regression parameters with known power and related estimation procedures. *Bull. Calcutta Statist. Ass.*, **9**, 19–34. [5.4]

Cherian, K.C. (1941). A bivariate correlated gamma-type distribution function. *J. Indian Math. Soc.*, **5**, 133–44. [10.10]

Courant, R. and Hilbert, D., (1953). *Methods of Mathematical Physics*, **1**. New York: Interscience Publishers. [7.2]

Cramér, H. (1946). *Mathematical Methods of Statistics*. Princeton: University Press. [9.5]

Daniels, H.E. (1954). Saddlepoint approximations in statistics. *Ann. Math. Statist.*, **25**, 631–50. [3.1]

David, F.N. and Fix, E. (1961). Rank correlation and regression in a non-normal surface. *Proc. 4th Berkeley Symp. Math. Statist. & Prob.*, **1**, 177–97. [10.10]

Dunnett, C.W. and Sobel, M. (1954). A bivariate generalization of Student's *t*-distribution, with tables for certain special cases. *Biometrika*, **41**, 153–69. [10.13]

Eagleson, G.K. (1964). Polynomial expansions of bivariate distributions. *Ann. Math. Statist.*, **35**, 1208–15. [9.1]

Edgeworth, F.Y. (1896). The compound law of error. *Phil. Mag.*, **41**, 207–15. [3.1, 3.2, 4.1]

Edgeworth, F.Y. (1905). The law of error. II. *Camb. Phil. Trans.*, **20**, 113–41. [3.1]

Edgeworth, F.Y. (1914). On the use of analytical geometry to present certain kinds of statistics. *J. R. Statist. Soc.*, **77**, 838–52. [4.1]

Edgeworth, F.Y. (1917). On the mathematical representation of statistical data. *J. R. Statist. Soc.*, **80**, 266–88. [3.1, 4.1]

Edwards, C.B. and Gurland, J. (1961). A class of distributions applicable to accidents. *J. Amer. Statist. Ass.*, **56**, 503–17. [10.4]

Erdélyi, A. *et al.* (1953). *Higher Transcendental Functions*. Vol. 2. New York: McGraw-Hill. [7.4]

Farlie, D.J.G. (1960). The performance of some correlation coefficients for a general bivariate distribution. *Biometrika*, **47**, 307–23. [9.2]

Farlie, D.J.G. (1961). The asymptotic efficiency of Daniels's generalized correlation coefficients. *J. R. Statist. Soc.*, B, **23**, 128–42. [7.5, 9.2]

Feller, W. (1966, 1968). *An Introduction to Probability Theory and its Applications*, **2**, **1** (3rd edn). New York: Wiley. [5.2, 9.1]

Fieller, E.C., Hartley, H.O. and Pearson, E.S. (1957). Tests for rank correlation coefficients. I. *Biometrika*, **44**, 470–81. [4.3e]

Fisher, R.A. (1940). The precision of discriminant functions. *Ann. Eugen.*, **10**, 422–9. [7.3]

Fréchet, M. (1951). Sur les tableaux de corrélation dont les marges sont données. Ann. Univ. Lyon, Section A, Series 3, **14**, 53–77. [4.3, 4.3a]

Fréchet, M. (1958). Remarques au sujet de la note précédente. *C.R. Acad. Sci., Paris*, **246**, 2719–20. [4.3a]

Galton, F. (1886). Family likeness in stature. With an appendix by J.D.H. Dickson. *Proc. Roy. Soc.*, **40**, 42–73. [1.1]

Gayen, A.K. (1951). The frequency distribution of the product-moment correlation coefficient in random samples of any size drawn from non-normal universes. *Biometrika*, **38**, 219–47. [3.4b]

Ghosh, M.N. (1955). Simultaneous tests of linear hypotheses. *Biometrika*, **42**, 441–9. [10.14]

Good, I.J. (1957). Saddle-point methods for the multinomial distribution. *Ann. Math. Statist.*, **28**, 861–81. [3.1]

Good, I.J. (1961). The multivariate saddlepoint method and chi-squared for the multinomial distribution. *Ann. Math. Statist.*, **32**, 535–48. [3.1]

Guldberg, A. (1934). On discontinuous frequency functions of two variables. *Skand. Aktuar.*, **17**, 89–117. [10.4]

Gumbel, E.J. (1958). Distributions a plusieurs variables dont les marges sont données. *C. R. Acad. Sci., Paris*, **246**, 2717–9. [9.2]

Gumbel, E.J. (1960). Bivariate exponential distributions. *J. Amer. Statist. Ass.*, **55**, 698–707. [9.2, 10.11]

Gumbel, E.J. (1961). Bivariate logistic distributions. *J. Amer. Statist. Ass.*, **56**, 335–49. [9.2]

Gumbel, E.J. and Mustafi, C.K. (1967). Some analytical properties of bivariate extremal distributions. *J. Amer. Statist. Ass.*, **62**, 569–88. [9.2]

Gupta, S.S. (1963a). Probability integrals of multivariate normal and multivariate *t*. *Ann. Math. Statist.*, **34**, 792–828. [10.13]

Gupta, S.S. (1963b). Bibliography on the multivariate normal integrals and related topics. *Ann. Math. Statist.*, **34**, 829–38. [10.13]

Haight, F.A. (1961). Index to the distributions of mathematical statistics. *J. Res. Nat. Bur. Stand.*, **65B**, 23–60. [1.1, Ch. 10]

Haight, F.A. (1967). *Handbook of the Poisson Distribution*. New York: Wiley. [10.3]

Haldane, J.B.S. (1949). A note on non-normal correlation. *Biometrika*, **36**, 467–8. [5.4]

Haley, D.C. (1952). Estimation of the dosage mortality relationship when the dose is subject to error. *Technical Report No. 15*, Applied Mathematics and Statistics Laboratory, Stanford Univ. [4.3c]

Hill, I.D. (1954). The distribution of the regression coefficient in samples from a non-normal population. *Biometrika*, **41**, 548–52. [10.16]

Hoeffding, W. (1940). Massstabinvariante Korrelations-theorie. *Schriften Math. Inst. Univ. Berlin*, **5**, 181–233. [8.2a]

Holgate, P. (1964). Estimation for the bivariate Poisson distribution. *Biometrika*, **51**, 241–5. [10.3]

Hopkins, J.W. and Clay, P.P.F. (1963). Some empirical distributions of bivariate T^2 and homoscedasticity, criterion M under unequal variance and leptokurtosis. *J. Amer. Statist. Ass.*, **58**, 1048–53. [9.4]

Isserlis, L. (1914). The application of solid hypergeometrical series to frequency distributions in space. *Phil. Mag.*, **28**, 379–403. [10.2]

Jogdeo, K. (1964). Nonparametric methods for regression. Mathematics Centre, Amsterdam, *Report S 330*. [5.4]

Jogdeo, K. (1968). Characterizations of independence in certain families of bivariate and multivariate distributions. *Ann. Math. Statist.*, **39**, 433–41. [6.2]

Johnson, N.L. (1949a). Systems of frequency curves generated by methods of translation. *Biometrika*, **36**, 149–76. [4.2a, 4.2c, 4.2e, 8.6]

Johnson, N.L. (1949b). Bivariate distributions based on simple translation systems. *Biometrika*, **36**, 297–304. [4.1, 4.2d, 4.2e]

Johnson, N.L. (1960). An approximation to the multinomial distribution: some properties and applications. *Biometrika*, **47**, 93–102. [10.9]

Jørgensen, N.R. (1916). *Undersøgelser over Frequensflader og Korrelation.* Copenhagen: Busck. [3.1, 5.3]

Kendall, M.G. (1949). Rank and product-moment correlation. *Biometrika*, **36**, 177–93. [3.1, 3.4b]

Kendall, M.G. (1957, 1968). *A Course in Multivariate Analysis.* London: Griffin. [10.9]

Kendall, M.G. and Stuart, A. (1969, 1967). *The Advanced Theory of Statistics.* Vol. 1 (3rd edn), Vol. 2 (2nd edn). London: Griffin. [2.2a, 3.1, 3.4a, 4.3b, 8.2a, 8.5b, 8.5c]

Khatri, C.G. (1959). On certain properties of power-series distributions. *Biometrika*, **46**, 486–90. [9.3, 10.5]

Kibble, W.F. (1941). A two-variate gamma type distribution. *Sankhyā*, **5**, 137–50. [10.10]

Konijn, H.S. (1956). On the power of certain tests for independence in bivariate populations. *Ann. Math. Statist.*, **27**, 300–23. (Corrections in *Ann. Math. Statist.*, **29**, 935–6). [5.4]

Konijn, H.S. (1957). A class of two-dimensional random variables and distribution functions. *Sankhyā*, **18**, 167–72. [5.2]

Konijn, H.S. (1959). Positive and negative dependence of two random variables. *Sankhyā*, **21**, 269–80. [4.3a]

Lancaster, H.O. (1957). Some properties of the bivariate normal distribution considered in the form of a contingency table. *Biometrika*, **44**, 289–92. [4.3b]

Lancaster, H.O. (1958). The structure of bivariate distributions. *Ann. Math. Statist.*, **29**, 719–36. [7.1, 7.2, 7.3, 7.5]

Lancaster, H.O. (1963). Correlations and canonical forms of bivariate distributions. *Ann. Math. Statist.*, **34**, 532–8. [7.3]

Lancaster, H.O. and Hamdan, M.A. (1964). Estimation of the correlation coefficient in contingency tables with possibly nonmetrical characters. *Psychometrika*, **29**, 383–91. [7.5]

Lehmann, E.L. (1966). Some concepts of dependence. *Ann. Math. Statist.*, **37**, 1137–53. [8.2a]

Leipnik, R. (1959). Integral equations, biorthonormal expansions, and noise. *J. Soc. Indust. Appl. Math.*, **7**, 6–30. [7.4]

Lundberg, O. (1940). *On Random Processes and their Application to Sickness and Accident Statistics.* Uppsala: Almqvist and Wiksell. [10.4]

McKay, A.T. (1934). Sampling from batches. *J. R. Statist. Soc.*, B, **1**, 207–216. [10.10]

Mardia, K.V. (1962). Multivariate Pareto distributions. *Ann. Math. Statist.*, 33, 1008–15. [10.12]

Mardia, K.V. (1967a). Some contributions to contingency-type bivariate distributions. *Biometrika*, **54**, 235–49. (Corrections in *Biometrika*, **55**, 597). [8.1, 8.2a, 8.4, 8.5c, 8.6, 8.7]

Mardia, K.V. (1967b). A non-parametric test for the bivariate two-sample location problem. *J.R. Statist. Soc.*, B, **29**, 320–42. [9.5]

Mardia, K.V. (1969). The performance of some tests of independence for contingency-type bivariate distributions. *Biometrika*, **56**, 449–51. [8.7]

Marshall, A.W. and Olkin, I. (1967). A multivariate exponential distribution. *J. Amer. Statist. Ass.*, **62**, 30–44. [10.11]

Maung, K. (1942). Measurement of association in a contingency table with special reference to the pigmentation of hair and eye colours of Scottish schoolchildren. *Ann. Eugen.*, **11**, 189–205. [7.3]

Mehr, C.B. (1967). Characterization of a class of second-order density functions. *J. Appl. Prob.*, **4**, 123–9. [7.4]

Mehr, C.B. (1968). An application of biorthonormal expansions in theory of stochastic processes. *J. R. Statist. Soc.*, B, **30**, 334–7. [7.4]

Mihram, G.A. and Hulquist, A.R. (1967). A bivariate warning-time/failure-time distribution. *J. Amer. Statist. Ass.*, **62**, 589–99. [10.10]

Moran, P.A.P. (1967). Testing for correlation between non-negative variates. *Biometrika*, **54**, 385–94. [10.10]

Morgenstern D. (1956). Einfache Beispiele zweidimensionaler Verteilungen. *Mitt. Math. Statist.*, **8**, 234–5. [8.2b, 9.2, 10.8]

Mosteller, F. (1968). Association and estimation in contingency tables. *J. Amer. Statist. Ass.*, **63**, 1–28. [8.1]

Narumi, S. (1923a,b). On the general forms of bivariate frequency distributions which are mathematically possible when regression and variation are subjected to limiting conditions. I, II. *Biometrika*, **15**, 77–88, 209–21. [6.1, 6.2]

Nataf, A. (1962). Détermination des distributions de probabilités dont les marges sont données. *C. R. Acad. Sci., Paris*, **255**, 42–3. [4.3]

Neyman, J. (1965). Certain chance mechanisms involving discrete distributions, 4–17. In *Classical and Contagious Discrete Distributions*, ed. G.P. Patil, London: Pergamon Press. [10.4]

Olkin, I. and Rubin, H. (1964). Multivariate beta distributions and independence properties of the Wishart distribution. *Ann. Math. Statist.*, **35**, 261–9. [10.9]

Ord, J.K. (1967a). On a system of discrete distributions. *Biometrika*, **54**, 649–56. [1.1]

Ord, J.K. (1967b). Graphical methods for a class of discrete distributions. *J. R. Statist. Soc.*, A, 130, 232–8. [1.1]

Patil, G.P. (1965). On multivariate generalized power series distribution and its application to the multinomial and negative multinomial, 183–94. In *Classical and Contagious Discrete Distributions*, ed. G.P. Patil. London: Pergamon Press. [9.3]

Patil, G.P. and Bildikar, S. (1967). Multivariate logarithmic series distribution as a probability model in population and community ecology and some of its statistical properties. *J. Amer. Statist. Ass.*, **62**, 655–74. [10.5]

Pearson, E.S. (1962). Frequency surfaces. Statistical Techniques Research Group, *Technical Report No. 49*. Princeton Univ. [1.1, 4.2c]

Pearson, K. (1923a). Notes on skew frequency surfaces. *Biometrika*, **15**, 222–30. [2.1, 5.3, 10.9]

Pearson, K. (1923b). On non-skew frequency surfaces. *Biometrika*, **15**, 231–44. [10.8, 10.13, 10.15]

Pearson, K. (1924a). On the moments of the hypergeometrical series. *Biometrika*, **16**, 157–60. [10.2]

Pearson, K. (1924b). On a certain double hypergeometric series and its representation by continuous frequency surfaces. *Biometrika*, **16**, 172–88. [10.2, 10.9, 10.13]

Pearson, K. (1925). The fifteen constant bivariate frequency surface. *Biometrika*, **17**, 268–313. [3.1, 3.2, 3.3, 3.4a]

Plackett, R.L. (1965). A class of bivariate distributions. *J. Amer. Statist. Ass.*, **60**, 516–22. [8.1, 8.4, 8.7]

Pretorius, S.J. (1930). Skew bivariate frequency surfaces, examined in the light of numerical illustrations. *Biometrika*, **22**, 109–223. [1.1, 3.1, 3.3, 3.4a, 4.1, 4.2c, 6.2, 9.6]

Rao, C.R. (1965). *Linear Statistical Inference and its Applications*. New York: Wiley. [10.8]

Rényi, A. (1959). On measures of dependence. *Acta Math., Budapest*, **10**, 441–51. [7.2]

Rhodes, E.C. (1923). On a certain skew correlation surface. *Biometrika*, **14**, 355–77. [2.1, 10.16]

Rice, S.O. (1944, 1945). Mathematical analysis of random noise. *Bell Syst. Tech. J.*, **23**, 282–332; **24**, 46–156. [10.10]

Rosenblatt, M. (1952). Remarks on a multivariate transformation. *Ann. Math. Statist.*, **23**, 470–2. [8.2c]

Sagrista, S.N. (1952). Sobre una generalizacion de las curvas de Pearson al caso bidimensional. *Trab. Estadist.*, **3**, 273–314. [2.1, 10.15]

Seshadri, V. and Patil, G.P. (1964). A characterization of a bivariate distribution by the marginal and the conditional distributions of the same component. *Ann. Inst. Statist. Math.*, **15**, 215–21. [10.3]

Sibuya, M. (1960). Bivariate extreme statistics. I. *Ann. Inst. Statist. Math.*, **11**, 195–210. [9.2]

Sibuya, M., Yoshimura, I. and Shimizu, R. (1964). Negative multinomial distribution. *Ann. Inst. Statist. Math.*, **16**, 409–26. [10.4]

Siddiqui, M.M. (1967). A bivariate *t* distribution. *Ann. Math. Statist.*, **38**, 162–6. [10.13]

Srivastava, A.B.L. (1960). The distribution of regression coefficients in samples from bivariate non-normal populations. I. Theoretical investigation. *Biometrika*, **47**, 61–8. [3.4b]

Steck, G.P. (1968). A note on contingency-type bivariate distributions. *Biometrika*, **55**, 262–4. [8.2a]

Steffensen, J.F. (1922). A correlation-formula. *Skand. Aktuar.*, **5**, 73–91. [5.1, 5.3]

Steffensen, J.F. (1941). On the coefficient of correlation for continuous distributions. *Skand. Aktuar.*, **24**, 1–12. (Correction in *Skand Aktuar.*, **24**, 232) [5.2]

Van der Stok, J.P. (1908). On the analysis of frequency curves according to a general method. *Ned. Akad. Wet. Proc.*, **10**, 799–817. [3.1]

Van Uven, M.J. (1925a, b; 1926, 1929). On treating skew correlation. *Ned. Akad. Wet. Proc.*, **28**, 797–811, 919–35; **29**, 580–90; **32**, 408–13. [4.3b]

Van Uven, M.J. (1947a,b; 1948a,b). Extension of Pearson's probability distributions to two variables. I-IV. *Ned. Akad. Wet. Proc.*, **50**, 1063–70, 1252–64; **51**, 41–52, 191–6. [2.2a, 2.2b, 2.2c]

Vere-Jones, D. (1967). The infinite divisibility of a bivariate gamma distribution. *Sankhyā*, A, **29**, 421–2. [10.10]

Wicksell, S.D. (1917a). The correlation function of Type A. *Medd. Lunds Astr. Obs.*, Series 2, No. 17. [3.1, 3.2, 3.3, 3.4a]

Wicksell, S.D. (1917b). The construction of the curves of equal frequency in case of type A correlation. *Svensk. Aktuar.*, **4**, 122–40. [3.1, 3.2]

Wicksell, S.D. (1917c). On logarithmic correlation, with an application to the distribution of ages at first marriage. *Svensk. Aktuar.*, **4**, 1–21. [4.2a]

Wicksell, S.D. (1933). On correlation functions of Type III. *Biometrika*, **25**, 121–33. [10.10]

Wilks, S.S. (1962). *Mathematical Statistics*, 2nd edn. New York: Wiley. [10.8]

Yuan, P.T. (1933). On the logarithmic frequency distribution and the semi-logarithmic correlation surface. *Ann. Math. Statist.*, **4**, 30–74. [4.2a]

SUBJECT INDEX